Alternative Energy

Alternative Energy

Volume 3

Neil Schlager and Jayne Weisblatt, editors

U·X·L

An imprint of Thomson Gale, a part of The Thomson Corporation

THOMSON

GALE

Detroit • New York • San Francisco • San Diego • New Haven, Conn. • Waterville, Maine • London • Munich

Alternative Energy

Neil Schlager and Jayne Weisblatt, Editors

Project Editor
Madeline S. Harris

Editorial
Luann Brennan, Marc Faeber, Kristine Krapp, Elizabeth Manar, Kim McGrath, Paul Lewon, Rebecca Parks, Heather Price, Lemma Shomali

Indexing Services
Factiva, a Dow Jones & Reuters Company

Rights and Acquisitions
Margaret Abendroth, Timothy Sisler

Imaging and Multimedia
Randy Bassett, Lezlie Light, Michael Logusz, Christine O'Bryan, Denay Wilding

Product Design
Jennifer Wahi

Composition
Evi Seoud, Mary Beth Trimper

Manufacturing
Wendy Blurton, Dorothy Maki

LIBRARY OF CONGRESS CATALOGING-IN-PUBLICATION DATA

Alternative energy / Neil Schlager and Jayne Weisblatt, editors.
    p. cm.
        Includes bibliographical references and index.
        ISBN 0-7876-9440-1 (set hardcover : alk. paper) –
        ISBN 0-7876-9439-8 (vol 1 : alk. paper) –
        ISBN 0-7876-9441-X (vol 2 : alk. paper) –
        ISBN 0-7876-9442-8 (vol 3 : alk. paper)
        1. Renewable energy sources. I. Schlager, Neil, 1966- II. Weisblatt, Jayne.

TJ808.A475 2006
333.79′4–dc22

2006003763

This title is also available as an e-book
ISBN 1-4414-0507-3
Contact your Thomson Gale sales representative for ordering information.

Printed in China
10 9 8 7 6 5 4 3 2

■■■ Contents

Introduction viii
Words to Know x
Overview xxi

CHAPTER 1: FOSSIL FUELS

Introduction: What are Fossil Fuels? 1
Petroleum 20
Natural Gas 30
Coal 38
Coal Gasification 44
Liquefied Petroleum Gas: Propane and Butane 46
Methanol 50
Methyl Tertiary-Butyl Ether 52
For More Information 54

CHAPTER 2: BIOENERGY

Introduction: What is Bioenergy? 57
Solid Biomass 69
Biodiesel 75
Vegetable Oil Fuels 80
Biogas 84
Ethanol and Other Alcohol Fuels 87
P-Series Fuels 92
For More Information 94

CHAPTER 3: GEOTHERMAL ENERGY

Introduction: What is Geothermal Energy? 97
Agricultural Applications 115
Aquacultural Applications 118

Geothermal Power Plants. 121
Geothermal Heating Applications 125
Industrial Applications. 130
For More Information 131

CHAPTER 4: HYDROGEN
Introduction: What is Hydrogen Energy? 133
Historical Overview 134
Producing Hydrogen 147
Using Hydrogen 150
Transporting Hydrogen 157
Distributing Hydrogen 159
Storing Hydrogen. 160
Impacts. 162
Future Technology. 165
For More Information 166

CHAPTER 5: NUCLEAR ENERGY
Introduction: What is Nuclear Energy?. 169
Historical Overview 171
How Nuclear Energy Works 184
Current and Future Technology 188
Benefits and Drawbacks. 192
Environmental Impact 202
Economic Impact. 204
Societal Impact. 206
Barriers to Implementation or Acceptance 207
For More Information 208

CHAPTER 6: SOLAR ENERGY
Introduction: What is Solar Energy? 209
Passive Solar Design. 222
Daylighting. 226
Transpired Solar Collectors 228
Solar Water Heating Systems. 230
Photovoltaic Cells 236
Dish Systems 243
Trough Systems 246
Solar Ponds. 248
Solar Towers. 253
Solar Furnaces 255
For More Information 258

CHAPTER 7: WATER ENERGY
Introduction: What is Water Energy? 261
Hydropower 275

Hydroelectricity 279
Ocean Thermal Energy Conversion. 290
Tidal Power 294
Ocean Wave Power 299
For More Information 302

CHAPTER 8: WIND ENERGY

Introduction: What is Wind Energy? 305
How Wind Energy Works 317
Current and Future Technology 321
Benefits and Drawbacks of Wind Energy 321
Wind Turbines. 324
Current and Potential Uses 329
Issues, Challenges, and Obstacles 334
For More Information 336

CHAPTER 9: ENERGY CONSERVATION AND EFFICIENCY

Introduction. 337
Climate Responsive Buildings 341
Green Building Materials. 347
Lighting 352
Energy Efficiency and Conservation in the Home 358
Transportation 364
Hybrid Vehicles 365
Leaving an Energy Footprint on the Earth 374
For More Information 377

CHAPTER 10: POSSIBLE FUTURE ENERGY SOURCES

Is Alternative Energy Enough? 380
Dreams of Free Energy 384
Perpetual Motion, an Energy Fraud and Scam 385
Advances in Electricity and Magnetism. 386
Zero Point Energy 390
Fusion. 395
Solar Power Satellites. 407
No Magic Bullets 408
For More Information 409

WHERE TO LEARN MORE

WHERE TO LEARN MORE xxix

INDEX xxxix

■■■

Introduction

Alternative Energy offers readers comprehensive and easy-to-use information on the development of alternative energy sources. Although the set focuses on new or emerging energy sources, such as geothermal power and solar energy, it also discusses existing energy sources such as those that rely on fossil fuels. Each volume begins with a general overview that presents the complex issues surrounding existing and potential energy sources. These include the increasing need for energy, the world's current dependence on nonrenewable sources of energy, the impact on the environment of current energy sources, and implications for the future. The overview will help readers place the new and alternative energy sources in perspective.

Each of the first eight chapters in the set covers a different energy source. These chapters each begin with an overview that defines the source, discusses its history and the scientists who developed it, and outlines the applications and technologies for using the source. Following the chapter overview, readers will find information about specific technologies in use and potential uses as well. Two additional chapters explore the need for conservation and the move toward more energy-efficient tools, building materials, and vehicles and the more theoretical (and even imaginary) energy sources that might become reality in the future.

ADDITIONAL FEATURES

Each volume of *Alternative Energy* includes the overview, a glossary called "Words to Know," a list of sources for more information, and an index. The set has 100 photos, charts, and illustrations to enliven the text, and sidebars provide additional facts and related information.

ACKNOWLEDGEMENTS

U•X•L would like to thank several individuals for their assistance with this set. At Schlager Group, Jayne Weisblatt and Neil Schlager oversaw the writing and editing of the set. Michael J. O'Neal, Amy Hackney Blackwell, and A. Petruso wrote the text for the volumes.

In addition, U•X•L editors would like to thank Dr. Peter Brimblecombe for his expert review of these volumes. Dr. Brimblecombe teaches courses on air pollution at the School of Environmental Sciences, University of East Anglia, United Kingdom. The editors also express their thanks for last minute contributions, review, and revisions to the final chapter on alternative and potential energy resources to Rory Clarke (physicist, CERN), Lee Wilmoth Lerner (electrical engineer and intern, NASA and the Fusion Research Laboratory at Auburn University), Larry Gilman (electrical engineer), and K. Lee Lerner (physicist and managing director, Lerner & Lerner, LLC).

COMMENTS AND SUGGESTIONS

We welcome your comments on *Alternative Energy* and suggestions for future editions of this work. Please write: Editors, *Alternative Energy*, U•X•L, 27500 Drake Rd., Farmington Hills, Michigan 48331-3535; call toll free: 1-800-877-4253; fax: 248-699-8097; or send e-mail via www.gale.com.

Words to Know

A

acid rain: Rain with a high concentration of sulfuric acid, which can damage cars, buildings, plants, and water supplies where it falls.

adobe: Bricks that are made from clay or earth, water, and straw, and dried in the sun.

alkane: A kind of hydrocarbon in which the molecules have the maximum possible number of hydrogen atoms and no double bonds.

anaerobic: Without air; in the absence of air or oxygen.

anemometer: A device used to measure wind speed.

anthracite: A hard, black coal that burns with little smoke.

aquaculture: The formal cultivation of fish or other aquatic life forms.

atomic number: The number of protons in the nucleus of an atom.

atomic weight: The combined number of an atom's protons and neutrons.

attenuator: A device that reduces the strength of an energy wave, such as sunlight.

B

balneology: The science of bathing in hot water.

barrel: A common unit of measurement of crude oil, equivalent to 42 U.S. gallons; barrels of oil per day, or BOPD, is a standard measurement of how much crude oil a well produces.

biodiesel: Diesel fuel made from vegetable oil.

bioenergy: Energy produced through the combustion of organic materials that are constantly being created, such as plants.

biofuel: A fuel made from organic materials that are constantly being created.

biomass: Organic materials that are constantly being created, such as plants.

bitumen: A black, viscous (oily) hydrocarbon substance left over from petroleum refining, often used to pave roads.

bituminous coal: Mid-grade coal that burns with a relatively high flame and smoke.

brine: Water that is very salty, such as the water found in the ocean.

British thermal unit (Btu or BTU): A measure of heat energy, equivalent to the amount of energy it takes to raise the temperature of one pound of water by one degree Fahrenheit.

butyl rubber: A synthetic rubber that does not easily tear. It is often used in hoses and inner tubes.

C

carbon sequestration: Storing the carbon emissions produced by coal-burning power plants so that pollutants are not released in the atmosphere.

catalyst: A substance that speeds up a chemical reaction or allows it to occur under different conditions than otherwise possible.

cauldron: A large metal pot.

CFC (chlorofluorocarbon): A chemical compound used as a refrigerant and propellant before being banned for fear it was destroying the ozone layer.

Clean Air Act: A U.S. law intended to reduce and control air pollution by setting emissions limits for utilities.

climate-responsive building: A building, or the process of constructing a building, using materials and techniques that take advantage of natural conditions to heat, cool, and light the building.

coal: A solid hydrocarbon found in the ground and formed from plant matter compressed for millions of years.

coke: A solid organic fuel made by burning off the volatile components of coal in the absence of air.

cold fusion: Nuclear fusion that occurs without high heat; also referred to as low energy nuclear reactions.

combustion: Burning.

compact fluorescent bulb: A lightbulb that saves energy as conventional fluorescent bulbs do, but that can be used in fixtures that normally take incandescent lightbulbs.

compressed: To make more dense so that a substance takes up less space.

conductive: A material that can transmit electrical energy.

convection: The circulation movement of a substance resulting from areas of different temperatures and/or densities.

core: The center of the Earth.

coriolis force: The movement of air currents to the right or left caused by Earth's rotation.

corrugated steel: Steel pieces that have parallel ridges and troughs.

critical mass: An amount of fissile material needed to produce an ongoing nuclear chain reaction.

criticality: The point at which a nuclear fission reaction is in controlled balance.

crude oil: The unrefined petroleum removed from an oil well.

crust: The outermost layer of the Earth.

curie: A unit of measurement that measures an amount of radiation.

current: The flow of electricity.

D

decay: The breakdown of a radioactive substance over time as its atoms spontaneously give off neutrons.

deciduous trees: Trees that shed their leaves in the fall and grow them in the spring. Such trees include maples and oaks.

decommission: To take a nuclear power plant out of operation.

dependent: To be reliant on something.

distillation: A process of separating or purifying a liquid by boiling the substance and then condensing the product.

distiller's grain: Grain left over from the process of distilling ethanol, which can be used as inexpensive high-protein animal feed.

drag: The slowing force of the wind as it strikes an object.

drag coefficient: A measurement of the drag produced when an object such as a car pushes its way through the air.

E

E85: A blend of 15 percent ethanol and 85 percent gasoline.

efficient: To get a task done without much waste.

electrolysis: A method of producing chemical energy by passing an electric current through a type of liquid.

electromagnetism: Magnetism developed by a current of electricity.

electron: A negatively charged particle that revolves around the nucleus in an atom.

embargo: Preventing the trade of a certain type of commodity.

emission: The release of substances into the atmosphere. These substances can be gases or particles.

emulsion: A liquid that contains many small droplets of a substance that cannot dissolve in the liquid, such as oil and water shaken together.

enrichment: The process of increasing the purity of a radioactive element such as uranium to make it suitable as nuclear fuel.

ethanol: An alcohol made from plant materials such as corn or sugar cane that can be used as fuel.

experimentation: Scientific tests, sometimes of a new idea.

F

feasible: To be possible; able to be accomplished or brought about.

feedstock: A substance used as a raw material in the creation of another substance.

field: An area that contains many underground reservoirs of petroleum or natural gas.

fissile: Term used to describe any radioactive material that can be used as fuel because its atoms can be split.

fission: Splitting of an atom.

flexible fuel vehicle (FFV): A vehicle that can run on a variety of fuel types without modification of the engine.

flow: The volume of water in a river or stream, usually expressed as gallons or cubic meters per unit of time, such as a minute or second.

fluorescent lightbulb: A lightbulb that produces light not with intense heat but by exciting the atoms in a phosphor coating inside the bulb.

fossil fuel: An organic fuel made through the compression and heating of plant matter over millions of years, such as coal, petroleum, and natural gas.

fusion: The process by which the nuclei of light atoms join, releasing energy.

G

gas: An air-like substance that expands to fill whatever container holds it, including natural gas and other gases commonly found with liquid petroleum.

gasification: A process of converting the energy from a solid, such as coal, into gas.

gasohol: A blend of gasoline and ethanol.

gasoline: Refined liquid petroleum most commonly used as fuel in internal combustion engines.

geothermal: Describing energy that is found in the hot spots under the Earth; describing energy that is made from heat.

geothermal reservoir: A pocket of hot water contained within the Earth's mantle.

global warming: A phenomenon in which the average temperature of the Earth rises, melting icecaps, raising sea levels, and causing other environmental problems.

gradient: A gradual change in something over a specific distance.

green building: Any building constructed with materials that require less energy to produce and that save energy during the building's operation.

greenhouse effect: A phenomenon in which gases in the Earth's atmosphere prevent the sun's radiation from being reflected back into space, raising the surface temperature of the Earth.

greenhouse gas: A gas, such as carbon dioxide or methane, that is added to the Earth's atmosphere by human actions. These gases trap heat and contribute to global warming.

H

halogen lamp: An incandescent lightbulb that produces more light because it produces more heat, but lasts longer because the filament is enclosed in quartz.

Heisenberg uncertainty principle: The principle that it is impossible to know simultaneously both the location and momentum of a subatomic particle.

heliostat: A mirror that reflects the sun in a constant direction.

hybrid vehicle: Any vehicle that is powered in a combination of two ways; usually refers to vehicles powered by an internal combustion engine and an electric motor.

hybridized: The bringing together of two different types of technology.

hydraulic energy: The kinetic energy contained in water.

hydrocarbon: A substance composed of the elements hydrogen and carbon, such as coal, petroleum, and natural gas.

hydroelectric: Describing electric energy made by the movement of water.

hydropower: Any form of power derived from water.

I

implement: To put something into practice.

incandescent lightbulb: A conventional lightbulb that produces light by heating a filament to high temperatures.

infrastructure: The framework that is necessary to the functioning of a structure; for example, roads and power lines form part of the infrastructure of a city.

inlet: An opening through which liquid enters a device, or place.

internal combustion engine: The type of engine in which the burning that generates power takes place inside the engine.

isotope: A "species" of an element whose nucleus contains more neutrons than other species of the same element.

K

kilowatt-hour: One kilowatt of electricity consumed over a one-hour period.

kinetic energy: The energy associated with movement, such as water that is in motion.

Kyoto Protocol: An international agreement among many nations setting limits on emissions of greenhouse gases; intended to slow or prevent global warming.

L

lava: Molten rock contained within the Earth that emerges from cracks in the Earth's crust, such as volcanoes.

lift: The aerodynamic force that operates perpendicular to the wind, owing to differences in air pressure on either side of a turbine blade.

lignite: A soft brown coal with visible traces of plant matter in it that burns with a great deal of smoke and produces less heat than anthracite or bituminous coal.

liquefaction: The process of turning a gas or solid into a liquid.

LNG (liquefied natural gas): Gas that has been turned into liquid through the application of pressure and cold.

LPG (liquefied petroleum gas): A gas, mainly propane or butane, that has been turned into liquid through the use of pressure and cold.

lumen: A measure of the amount of light, defined as the amount of light produced by one candle.

M

magma: Liquid rock within the mantle.

magnetic levitation: The process of using the attractive and repulsive forces of magnetism to move objects such as trains.

mantle: The layer of the Earth between the core and the crust.

mechanical energy: The energy output of tools or machinery.

meltdown: Term used to refer to the possibility that a nuclear reactor could become so overheated that it would melt into the earth below.

mica: A type of shiny silica mineral usually found in certain types of rocks.

modular: An object which can be easily arranged, rearranged, replaced, or interchanged with similar objects.

mousse: A frothy mixture of oil and seawater in the area where an oil spill has occurred.

N

nacelle: The part of a wind turbine that houses the gearbox, generator, and other components.

natural gas: A gaseous hydrocarbon commonly found with petroleum.

negligible: To be so small as to be insignificant.

neutron: A particle with no electrical charge found in the nucleus of most atoms.

NGL (natural gas liquid): The liquid form of gases commonly found with natural gas, such as propane, butane, and ethane.

nonrenewable: To be limited in quantity and unable to be replaced.

nucleus: The center of an atom, containing protons and in the case of most elements, neutrons.

O

ocean thermal energy conversion (OTEC): The process of converting the heat contained in the oceans' water into electrical energy.

octane rating: The measure of how much a fuel can be compressed before it spontaneously ignites.

off-peak: Describing period of time when energy is being delivered at well below the maximum amount of demand, often nighttime.

oil: Liquid petroleum; a substance refined from petroleum used as a lubricant.

organic: Related to or derived from living matter, such as plants or animals; composed mainly of carbon atoms.

overburden: The dirt and rocks covering a deposit of coal or other fossil fuel.

oxygenate: A substance that increases the oxygen level in another substance.

ozone: A molecule consisting of three atoms of oxygen, naturally produced in the Earth's atmosphere; ozone is toxic to humans.

P

parabolic: Shaped like a parabola, which is a certain type of curve.

paraffin: A kind of alkane hydrocarbon that exists as a white, waxy solid at room temperature and can be used as fuel or as a wax for purposes such as sealing jars or making candles.

passive: A device that takes advantage of the sun's heat but does not use an additional source of energy.

peat: A brown substance composed of compressed plant matter and found in boggy areas; peat can be used as fuel itself, or turns into coal if compressed for long enough.

perpetual motion: The power of a machine to run indefinitely without any energy input.

petrochemicals: Chemical compounds that form in rocks, such as petroleum and coal.

petrodiesel: Diesel fuel made from petroleum.

petroleum: Liquid hydrocarbon found underground that can be refined into gasoline, diesel fuel, oils, kerosene, and other products.

pile: A mass of radioactive material in a nuclear reactor.

plutonium: A highly toxic element that can be used as fuel in nuclear reactors.

polymer: A compound, either synthetic or natural, that is made of many large molecules. These molecules are made from smaller, identical molecules that are chemically bonded.

pristine: Not changed by human hands; in its original condition.

productivity: The output of labor per amount of work.

proponent: Someone who supports an idea or cause.

proton: A positively charged particle found in the nucleus of an atom.

R

radioactive: Term used to describe any substance that decays over time by giving off subatomic particles such as neutrons.

RFG (reformulated gasoline):   Gasoline that has an oxygenate or other additive added to it to decrease emissions and improve performance.

rem: An abbreviation for "roentgen equivalent man," referring to a dose of radiation that will cause the same biological effect (on a "man") as one roentgen of X-rays or gamma rays.

reservoir: A geologic formation that can contain liquid petroleum and natural gas.

reservoir rock: Porous rock, such as limestone or sandstone, that can hold accumulations of petroleum or natural gas.

retrofit: To change something, like a home, after it is built.

rotor: The hub to which the blades of a wind turbine are connected; sometimes used to refer to the rotor itself and the blades as a single unit.

S

scupper: An opening that allows a liquid to drain.

seam: A deposit of coal in the ground.

sedimentary rock: A rock formed through years of minerals accumulating and being compressed.

seismology: The study of movement within the earth, such as earthquakes and the eruption of volcanoes.

sick building syndrome: The tendency of buildings that are poorly ventilated, lighted, and humidified, and that are made with certain synthetic materials to cause the occupants to feel ill.

smog: Air pollution composed of particles mixed with smoke, fog, or haze in the air.

stall: The loss of lift that occurs when a wing presents too steep an angle to the wind and low pressure along the upper surface of the wing decreases.

strip mining: A form of mining that involves removing earth and rocks by bulldozer to retrieve the minerals beneath them.

stored energy: The energy contained in water that is stored in a tank or held back behind a dam in a reservoir.

subsidence: The collapse of earth above an empty mine, resulting in a damaged landscape.

surcharge: An additional charge over and above the original cost.

superconductivity: The disappearance of electrical resistance in a substance such as some metals at very low temperatures.

T

thermal energy: Any form of energy in the form of heat; used in reference to heat in the oceans' waters.

thermal gradient: The differences in temperature between different layers of the oceans.

thermal mass: The measure of the amount of heat a substance can hold.

thermodynamics: The branch of physics that deals with the mechanical actions or relations of heat.

tokamak: An acronym for the Russian-built toroidal magnetic chamber, a device for containing a fusion reaction.

transitioning: Changing from one position or state to another.

transparent: So clear that light can pass through without distortion.

trap: A reservoir or area within Earth's crust made of nonporous rock that can contain liquids or gases, such as water, petroleum, and natural gas.

trawler: A large commercial fishing boat.

Trombé wall: An exterior wall that conserves energy by trapping heat between glazing and a thermal mass, then venting it into the living area.

turbine: A device that spins to produce electricity.

U

uranium: A heavy element that is the chief source of fuel for nuclear reactors.

V

viable: To be possible; to be able to grow or develop.

voltage: Electric potential that is measured in volts.

W

wind farm: A group of wind turbines that provide electricity for commercial uses.

work: The conversion of one form of energy into another, such as the conversion of the kinetic energy of water into mechanical energy used to perform a task.

Z

zero point energy: The energy contained in electromagnetic fluctuations that remains in a vacuum, even when the temperature has been reduced to very low levels.

Overview

In the technological world of the twenty-first century, few people can truly imagine the challenges faced by prehistoric people as they tried to cope with their natural environment. Thousands of years ago life was a daily struggle to find, store, and cook food, stay warm and clothed, and generally survive to an "old age" equal to that of most of today's college students. A common image of prehistoric life is that of dirty and ill-clad people huddled around a smoky campfire outside a cave in an ongoing effort to stay warm and dry and to stop the rumbling in their bellies.

The "caves" of the twenty-first century are a little cozier. The typical person, at least in more developed countries, wakes up each morning in a reasonably comfortable house because the gas, propane, or electric heating system (or electric air-conditioner) has operated automatically overnight. A warm shower awaits because of hot water heaters powered by electricity or natural gas, and hair dries quickly (and stylishly) under an electric hair dryer. An electric iron takes the wrinkles out of the clean shirt that sat overnight in the electric clothes dryer. Milk for a morning bowl of cereal remains fresh in an electric refrigerator, and it costs pennies per bowl thanks to electrically powered milking operations on modern dairy farms. The person then goes to the garage (after turning off all the electric lights in the house), hits the electric garage door opener, and gets into his or her gasoline-powered car for the drive to work—perhaps in an office building that consumes power for lighting, heating and air-conditioning, copiers, coffeemakers, and computers. Later, an electric, propane, or natural gas stove is used to cook dinner. Later still, an electric popcorn popper provides a snack as the person watches an electric television or reads under the warm glow of electric light bulbs— after perhaps turning up the heat because the house is a little chilly.

CATASTROPHE AHEAD?

Most people take these modern conveniences for granted. Few people give much thought to them, at least until there is a power outage or prices rise sharply, as they did for gasoline in the United States in the summer and fall of 2005. Many scientists, environmentalists, and concerned members of the public, though, believe that these conveniences have been taken too much for granted. Some believe that the modern reliance on fossil fuels—fuels such as natural gas, gasoline, propane, and coal that are processed from materials mined from the earth—has set the Earth on a collision course with disaster in the twenty-first century. Their belief is that the human community is simply burning too much fuel and that the consequences of doing so will be dire (terrible). Some of their concerns include the following:

- Too much money is spent on fossil fuels. In the United States, over $1 billion is spent every day to power the country's cars and trucks.
- Much of the supply of fossil fuels, particularly petroleum, comes from areas of the world that may be unstable. The U.S. fuel supply could be cut off without warning by a foreign government. Many nations that import all or most of their petroleum feel as if they are hostages to the nations that control the world's petroleum supplies.
- Drilling for oil and mining coal can do damage to the landscape that is impossible to repair.
- Reserves of coals and especially oil are limited, and eventually supplies will run out. In the meantime, the cost of such fuels will rise dramatically as it becomes more and more difficult to find and extract them.
- Transporting petroleum in massive tankers at sea heightens the risk of oil spills, causing damage to the marine and coastal environments.

Furthermore, to provide heat and electricity, fossil fuels have to be burned, and this burning gives rise to a host of problems. It releases pollutants in the form of carbon dioxide and sulfur into the air, fouling the atmosphere and causing "brown clouds" over cities. These pollutants can increase health problems such as lung disease. They may also contribute to a phenomenon called "global warming." This term refers to the theory that average temperatures across the globe will increase as "greenhouse gases" such as carbon dioxide trap the sun's heat (as a greenhouse does) in the atmosphere and warm it. Global warming, in turn, can melt glaciers and the polar ice caps, raising sea levels with damaging effects on coastal cities and small island nations. It may also cause climate changes, crop failures, and more unpredictable weather patterns.

Some scientists do not believe that global warming even exists or that its consequences will be catastrophic. Some note that throughout history, the world's average temperatures have risen and fallen. Some do not find the scientific data about temperature, glacial melting, rising sea levels, and unpredictable weather totally believable. While the debate continues, scientists struggle to learn more about the effects of human activity on the environment. At the same time, governments struggle to maintain a balance between economic development and its possible effects on the environment.

WHAT TO DO?

These problems began to become more serious after the Industrial Revolution of the nineteenth century. Until that time people depended on other sources of power. Of course, they burned coal or wood in fireplaces and stoves, but they also relied on the power of the sun, the wind, and river currents to accomplish much of their work. The Industrial Revolution changed that. Now, coal was being burned in vast amounts to power factories and steam engines as the economies of Europe and North America grew and developed. Later, more efficient electricity became the preferred power source, but coal still had to be burned to produce electricity in large power plants. Then in 1886 the first internal combustion engine was developed and used in an automobile. Within a few decades there was a demand for gasoline to power these engines. By 1929 the number of cars in the United States had grown to twenty-three million, and in the quarter-century between 1904 and 1929, the number of trucks grew from just seven hundred to 3.4 million.

At the same time technological advances improved life in the home. In 1920, for example, the United States produced a total of five thousand refrigerators. Just ten years later the number had grown to one million per year. These and many other industrial and consumer developments required vast and growing amounts of fuel. Compounding the problem in the twenty-first century is that other nations of the world, such as China and India, have started to develop more modern industrialized economies powered by fossil fuels.

By the end of World War II in 1945, scientists were beginning to imagine a world powered by fuel that was cheap, clean, and inexhaustible (unable to be used up). During the war the United States had unleashed the power of the atom to create the atomic bomb. Scientists believed that the atom could be used for peaceful purposes in nuclear power plants. They even envisioned (imagined) a day when homes could be powered by their own tiny nuclear power generators. This dream proved to be just that. While some four hundred nuclear power plants worldwide provide about 16 percent of the world's electricity, building such plants is an enormously expensive technical feat. Moreover, nuclear power plants produce spent fuel that is dangerous and not easily disposed of. The public fears that an accident at such a plant could release deadly radiation that would have disastrous effects on the surrounding area. Nuclear power has strong defenders, but it is not cheap, and safety concerns sometimes make it unpopular.

The dream of a fuel source that is safe, plentiful, clean, and inexpensive, however, lives on. The awareness of the need for such alternative fuel sources became greater in the 1970s, when the oil-exporting countries of the Middle East stopped shipments of oil to the United States and its allies. This situation (an embargo) caused fuel shortages and rapidly rising prices at the gas pump. In the decades that followed, gasoline again became plentiful and relatively inexpensive, but the oil embargo served as a wakeup call for many people. In addition, during these years people worldwide grew concerned about pollution, industrialization, and damage to the environment. Accordingly, efforts were intensified to find and develop alternative sources of energy.

ALTERNATIVE ENERGY: BACK TO THE FUTURE

Some of these alternative fuel sources are by no means new. For centuries people have harnessed the power of running water for a variety of needs, particularly for agriculture (farming). Water wheels were constructed in the Middle East, Greece, and China thousands of years ago, and they were common fixtures on the farms of Europe by the Middle Ages. In the early twenty-first century hydroelectric dams, which generate electricity from the power of rivers, provide about 9 percent of the electricity in the

United States. Worldwide, there are about 40,000 such dams. In some countries, such as Norway, hydroelectric dams provide virtually 100 percent of the nation's electrical needs. Scientists, though, express concerns about the impact such dams have on the natural environment.

Water can provide power in other ways. Scientists have been attempting to harness the enormous power contained in ocean waves, tides, and currents. Furthermore, they note that the oceans absorb enormous amounts of energy from the sun, and they hope someday to be able to tap into that energy for human needs. Technical problems continue to occur. It remains likely that ocean power will serve only to supplement (add to) existing power sources in the near future.

Another source of energy that is not new is solar power. For centuries, people have used the heat of the sun to warm houses, dry laundry, and preserve food. In the twenty-first century such "passive" uses of the sun's rays have been supplemented with photovoltaic devices that convert the energy of the sun into electricity. Solar power, though, is limited geographically to regions of the Earth where sunshine is plentiful.

Another old source of heat is geothermal power, referring to the heat that seeps out of the earth in places such as hot springs. In the past this heat was used directly, but in the modern world it is also used indirectly to produce electricity. In 1999 over 8,000 megawatts (that is, 8,000 million watts) of electricity were produced by about 250 geothermal power plants in twenty-two countries around the world. That same year the United States produced nearly 3,000 megawatts of geothermal electricity, more than twice the amount of power generated by wind and solar power. Geothermal power, though, is restricted by the limited number of suitable sites for tapping it.

Finally, wind power is getting a closer look. For centuries people have harnessed the power of the wind to turn windmills, using the energy to accomplish work. In the United States, wind-operated turbines produce just 0.4 percent of the nation's energy needs. However, wind experts believe that a realistic goal is for wind to supply 20 percent of the nation's electricity requirements by 2020. Worldwide, wind supplies enough power for about nine million homes. Its future development, though, is hampered by limitations on the number of sites with enough wind and by concerns about large numbers of unsightly wind turbines marring the landscape.

ALTERNATIVE ENERGY: FORWARD TO THE FUTURE

While some forms of modern alternative energy sources are really developments of long-existing technologies, others are genuinely new, though scientists have been exploring even some of these for up to hundreds of years. One, called bioenergy, refers to the burning of biological materials that otherwise might have just been thrown away or never grown in the first place. These include animal waste, garbage, straw, wood by-products, charcoal, dried plants, nutshells, and the material left over after the processing of certain foods, such as sugar and orange juice. Bioenergy also includes methane gas given off by garbage as it decomposes or rots. Fuels made from vegetable oils can be used to power engines, such as those in cars and trucks. Biofuels are generally cleaner than fossil fuels, so they do not pollute as much, and they are renewable. They remain expensive, and amassing significant amounts of biofuels requires a large commitment of agricultural resources such as farmland.

Nothing is sophisticated about burning garbage. A more sophisticated modern alternative is hydrogen, the most abundant element in the universe. Hydrogen in its pure form is extremely flammable. The problem with using hydrogen as a fuel is separating hydrogen molecules from the other elements to which it readily bonds, such as oxygen (hydrogen and oxygen combine to form water). Hydrogen can be used in fuel cells, where water is broken down into its elements. The hydrogen becomes fuel, while the "waste product" is oxygen. Many scientists regard hydrogen fuel cells as the "fuel of the future," believing that it will provide clean, safe, renewable fuel to power homes, office buildings, and even cars and trucks. However, fuel cells are expensive. As of 2002 a fuel cell could cost anywhere from $500 to $2,500 per kilowatt produced. Engines that burn gasoline cost only about $30 to $35 for the same amount of energy.

All of these power sources have high costs, both for the fuel and for the technology needed to use it. The real dreamers among energy researchers are those who envision a future powered by a fuel that is not only clean, safe, and renewable but essentially free. Many scientists believe that such fuel alternatives are impossible, at least for the foreseeable future. Others, though, work in laboratories around the world to harness more theoretical sources of energy. Some of their work has a "science fiction" quality, but these scientists point out that a few hundred years ago the airplane was science fiction.

One of these energy sources is magnetism, already used to power magnetic levitation ("maglev") trains in Japan and Germany. Another is perpetual motion, the movement of a machine that produces energy without requiring energy to be put into the system. Most scientists, though, dismiss perpetual motion as a violation of the laws of physics. Other scientists are investigating so-called zero-point energy, or the energy that surrounds all matter and can even be found in the vacuum of space. But perhaps the most sought-after source of energy for investigators is cold fusion, a nuclear reaction using "heavy hydrogen," an abundant element in seawater, as fuel. With cold fusion, power could be produced literally from a bucket of water. So far, no one has been able to produce it, though some scientists claim to have come very close.

None of these energy sources is a complete cure for the world's energy woes. Most will continue to serve as supplements to conventional fossil fuel burning for decades to come. But with the commitment of research dollars, it is possible that future generations will be able to generate all their power needs in ways that scientists have not even yet imagined. The first step begins with understanding fossil fuels, the energy they provide, the problems they cause, and what it may take to replace them.

Water Energy

INTRODUCTION: WHAT IS WATER ENERGY?

Water energy is energy derived from the power of water, most often its motion. Energy sources using water have been around for thousands of years in the form of water clocks and waterwheels. A more recent innovation has been hydroelectricity, or the electricity produced by the flow of water over dams. In the twenty-first century scientists are developing water-based applications ranging from tidal power to thermal power.

Historical overview

The history of water energy is almost as old as the history of human civilization itself, making it the first form of "alternative energy" people employed. Many centuries ago the ancient Egyptians devised water clocks, whose wheels were turned by the flow of water. The Egyptians and Syrians also used a device called a *noria,* a waterwheel with buckets attached, that was used to raise water out of the Nile River for use on their crops. Two thousand years ago the ancient Greeks built waterwheels to crush grapes and grind grains. At roughly the same time, the Chinese were using waterwheels to operate bellows used in the casting of iron tools such as farm implements.

The ancient Romans were especially skilled at managing water. In fact, the English word *plumber* comes from the Latin word *plumbum,* meaning "lead," referring to the lead pipes used in plumbing and reflected in the symbol for lead in the periodic table of elements, Pb. The Romans built water-carrying structures called aqueducts to channel water from natural sources to canals, where the water's energy could be harnessed by waterwheels. Near Arles in what is now southern France, for example, the

Words to Know

Flow The volume of water in a river or stream, usually expressed as gallons or cubic meters per unit of time, such as a minute or second.

Hydraulic energy The kinetic energy contained in water.

Hydropower Any form of power derived from water.

Kinetic energy The energy contained in any fluid mass, such as water, that is in motion.

Mechanical energy The energy output of tools or machinery.

Ocean thermal energy conversion (OTEC) The process of converting the heat contained in the oceans' water into electrical energy.

Stored energy The energy contained in water that is stored in a tank or held back behind a dam in a reservoir.

Thermal energy Any form of energy in the form of heat; used in reference to heat in the oceans' waters.

Thermal gradient The differences in temperature between different layers of the oceans.

Work The conversion of one form of energy into another, such as the conversion of the kinetic energy of water into mechanical energy used to perform a task.

Romans built a massive grain mill powered by sixteen water-wheels.

In the centuries that followed, until fossil fuels became the preferred power source during the industrial revolution of the nineteenth century, farmers continued to take advantage of the currents in rivers and streams for a variety of agricultural purposes, including grinding grain and pumping water for irrigation (watering crops). An English manuscript called the *Domesday Book,* written in 1086, listed 5,624 waterwheel-driven mills south of the Trent River in England, one mill for every four hundred people.

Farmers, though, were not the only ones to use waterwheels. Early factories, especially in Great Britain and in the American Northeast, relied heavily on water power as well because of the large number of rivers and streams in the British Isles and in such states as Massachusetts, Connecticut, and New York. In these examples, rivers often powered such enterprises as sawmills, but the textile industry, in particular, used water to power the "Spinning Jenny," a cotton-spinning machine for making cloth. In 1769 English inventor and industrialist Richard Arkwright (1732–1792) patented a water-powered textile loom for spinning cotton (originally meant to be powered by horses) that revolutionized the textile industry.

The result over the next half-century was a boom in the textile industry, both in Britain and, later, in the United States. One of the pioneers in this effort was a New England businessman, Francis Cabot Lowell (1775–1817). In the early nineteenth century Lowell imported British technology to the Charles River in Waltham, Massachusetts, where he and other business owners built textile mills powered by the river. Later, Waltham's mill owners, needing more power than the Charles could supply, moved to an area north of Boston. Here they created the industrial town of Lowell, Massachusetts, almost entirely around water power. Soon, textile mills were able to produce millions of yards of cloth, thanks largely to water power.

The major problem with early waterwheels, though, was that they could not store power for later use, nor could they easily distribute power to several users. This disadvantage was overcome by the development of hydroelectricity (though modern waterwheels can also produce electricity). Hydroelectric dams, unlike waterwheels, do not depend entirely on the rate of flow of the water in a river or stream. Moreover, by producing electricity, power can be stored and distributed to more than one user in a community.

The city of Hama in Syria is famous for its ancient water wheels, or *noria*, on the Orontes River. © *Elio Ciol/ Corbis*.

This is a traditional horizontal *noria* water wheel. The water comes out of the well on a wheel carrying pitchers, which then supplies the irrigation network. © *Marc Garanger/Corbis.*

Hydroelectricity was first used in 1880, when the Wolverine Chair Factory began producing hydroelectric power for its own use in its Grand Rapids, Michigan, plant (perhaps it is no accident that the city had the word *Rapids* in its name). The first hydroelectric plant whose power went to multiple customers began operation on September 30, 1882, on the Fox River near Appleton, Wisconsin. Major improvements in hydroelectric power generation were made by Lester Allan Pelton (1829–1908), an inventor who is sometimes called the "father of hydroelectric energy." Sometime in the late 1870s Pelton developed the Pelton wheel, a new, more efficient design for turbines that powered hydroelectric plants. A later design, developed by Eric Crewdson in 1920 and called the turgo impulse wheel, improved on the efficiency of Pelton's design. Because of these improvements, more and more electrical needs in the United States were being met by hydroelectric power.

The water in rivers and streams, though, is not the only water in motion. The oceans move too, and in the late twentieth and early twenty-first century, efforts have been launched to tap the power contained in the oceans' tides, waves, and currents. Fundamentally,

Richard Arkwright

Richard Arkwright, the youngest of thirteen children, began his career as a barber's apprentice. He wanted to run his own company, so he decided to become a wig maker. He spent the early part of his career traveling through England collecting discarded hair he could use to make wigs.

After Arkwright became involved in the textile industry in the 1760s, he built many profitable mills in England, Wales, and Scotland. When he died, he was worth nearly a million dollars, an enormous fortune in the late eighteenth century. In 1786 he was knighted by England's King George III.

But like many industrialists of the time, Arkwright built his fortune on the backs of his workers, who toiled from 6:00 in the morning to 7:00 in the evening. Among his 1,900 employees, two-thirds were children. While many other mill owners employed children as young as five, Arkwright was slightly enlightened for his time: he did not hire children under the age of six. Nor would he hire anyone over the age of forty.

though, these sources of power are little different from the power provided by rivers and streams. The water is moving, so the challenge for engineers is to devise ways to convert that motion into electricity. While strides have been made, the practical use of these power sources is still in the beginning stages.

Tidal power for electrical generation is relatively new. Currently, only one tidal power-generating station has been built and is in use. This plant is located at the mouth of the La Rance River along France's northern coast. The plant was built in 1966 and provides 240 megawatts, or 240 million watts, of electricity. There is a 20-megawatt experimental station in Nova Scotia, Canada, and Russia has a 0.4-megawatt station near the city of Murmansk. Other promising sites include the Severn River in western England, Cook Inlet in Alaska, and the White Sea in Russia.

Waves and ocean currents, like the tides, contain enormous amounts of energy, as any swimmer who has been pelted by a wave or swept along on an ocean current knows. The first patent for a wave power machine that would function much like a waterwheel in powering grain mills and sawmills was filed in France in 1799, although there is no evidence that the device was ever built.

The Pelton Wheel

Lester Allan Pelton (1829–1908) was born in Ohio but migrated to California during the gold rush of the late 1840s. In the 1870s he conceived the design for the Pelton wheel. He tested a prototype in 1879 and received a patent for the design in 1889.

Before the Pelton wheel, the most common type of turbine was the reaction turbine, which came equipped either with flat paddles or with cups or buckets. In either case, the water came straight at the paddle or bucket. As the water struck it, it pushed the paddle or bucket, thus turning the wheel. The Pelton wheel was the result of an accident. Pelton was watching a spinning water turbine. The key that held the wheel onto the shaft slipped out of place so that the wheel tilted. Instead of hitting the paddles on the waterwheel directly in the center, the water hit near the edge and was diverted to flow in a half-circle. To Pelton's surprise, the wheel actually began to spin faster.

The turgo turbine was developed in 1919 and represented an improvement in the Pelton wheel. It is less expensive to make and can handle a greater flow of water, so a smaller turgo turbine can generate the same amount of power as a larger Pelton wheel.

One of the first important developments for harnessing this power took place in 1974, when a British engineer named Stephen Salter invented a device called a "duck." This was a hydraulic mechanism that converted wave power into electricity, but this is only one of many ingenious innovations that scientists and engineers have developed. In the years that followed, scientists and engineers sought ways to transform innovations like the duck into a working wave power-generating station. Their efforts were finally successful in 2000, when the United Kingdom opened the first such station on the island of Islay, off the coast of Scotland. This station is called the Limpet 500, which stands for Land-Installed Marine-Powered Energy Transformer. The number 500 refers to the 500 kilowatts of electricity it feeds into the United Kingdom's power grid.

The world's oceans are also the source of thermal energy, or the heat that oceans absorb from the sun. The word *thermal* comes

Georges Claude

Georges Claude (1870–1960) may have built the first system for harnessing the thermal energy in oceans, but his impact as a scientist was probably much greater in a way that is glaringly obvious every day (or every night) in just about every city and town throughout the developed world.

As a young engineer, chemist, and inventor, Claude turned his attention to the inert gases. He discovered that passing an electrical current through cylinders filled with inert gases such as neon produces colored light. In other words, Claude was the inventor of the neon sign, which he first demonstrated in Paris in 1910. The first neon signs arrived in the United States when he sold two of them to a Packard automobile dealership in Los Angeles in 1923.

from a Greek word, *therme,* meaning "heat," and is related to another Greek word, *thermos,* meaning "hot."

The first scientist to propose that the thermal energy of the oceans could be tapped for human needs was a French physicist named Jacques Arsene d'Arsonval (1851–1940) in 1881. D'Arsonval may very well have gotten the idea, though, from author Jules Verne (1828–1905), who imagined the use of ocean temperature differences to produce electricity in his novel *Twenty Thousand Leagues under the Sea* in 1870. In 1930 one of d'Arsonval's students, Georges Claude, built the first-ever system for doing so off the coast of Cuba. The system he built generated 22 kilowatts, or 22,000 watts, of electricity. However, this it represented a net power loss, because it actually took more power to run the system than it was able to generate. Then in 1974 the Natural Energy Laboratory of Hawaii Authority (NELHA) was formed. In 1979 NELHA successfully demonstrated a plant that produced more energy than it consumed (50-kilowatts gross; 15-kilowatts net). In 1981 Japan built a system that produced 31.5 kilowatts of net power. In 1993 NELHA set a record when it produced a net power of 50 kilowatts in a demonstration.

How water energy works

To understand fully the nature of water energy, two terms have to be defined more precisely: energy and work. In everyday use, the word energy often refers to a substance, such as gasoline, coal, or natural gas. Strictly speaking, though, these substances are not energy; they are just chemical substances. Their energy is locked inside their chemical bonds, and it has to be released by burning them. What makes these substances useful is that they contain a lot of energy that can easily be released through combustion (burning).

Put differently, these substances can do a great deal of work, but scientists define work in their own peculiar way. To most people, "work" means something like a chore or job, such as mowing the lawn. To a scientist, though, "work" refers to the process of converting one form of energy into another, such as converting the chemical energy of natural gas into heat used to boil water or heat a house. Scientists usually measure energy output in terms of the amount of work that can be done with it. For example, the calorie, used most often in discussions of diet, exercise, and weight, is actually a unit that measures a form of work. A more commonly used unit of work among scientists is the joule. The joule is part of the metric system units, and it is used to measure heat, electric energy, and the energy of motion.

To produce energy, though, it is not always necessary to burn something. When cleaning up after dinner, a family's first task is to rinse off the dishes, pots, and pans, using water from the kitchen faucet. What rinses the dishes, though, is not the water from the faucet by itself so much as it is the energy contained in the running water. This type of energy is called kinetic energy. The word *kinetic* comes from a Greek word, *kinesis,* which means "motion," so kinetic energy is the energy contained in a body of water when it is in motion. In discussions of water energy, sometimes the term *hydraulic energy* is used instead of kinetic energy. The word *hydraulic* is derived from *hydro,* the Greek word for "water." In this context, kinetic energy and hydraulic energy refer to the same thing.

To put water to work, then, the water has to be in motion. The best way to put large amounts of water in motion is to let gravity do the work. Streams and rivers, for example, flow because the water in them is moving downhill, even if only slightly, following the downward pull of gravity. In a home, water flows "downhill" because a city's water is stored in large elevated tanks, where it contains stored energy. When a homeowner opens a faucet, the water flows in a downward direction from the tank through the city's water pipes and out the faucet, where it carries enough kinetic energy to knock food remnants off dirty dinner dishes. Helping out is the sheer weight of the water, which pushes it down through the city's water pipes.

Scientists measure how much work a body of water can do using flow, which is simply the volume of water measured in, for example, gallons or liters per second or minute. This is just common sense. A homeowner who wants to rinse off a dirty porch uses a hose, not a squirt gun, because the flow from the hose is much greater than the flow from a squirt gun, so the water can do more work in a given period of time. A squirt gun might work, but the job would take a very long time.

This, then, is the basic science behind kinetic energy. Water flowing downhill, pulled by gravity, contains kinetic energy. A tool such as a waterwheel can be used to convert this kinetic energy into mechanical energy, which can then be harnessed to perform a task, such as grinding grain, sawing lumber, or running a textile loom. Or the kinetic energy can be transformed into electricity, which can be stored and distributed to many different users.

Current and future technology

The moon in large part is responsible for another type of energy that water can provide: tidal power. Every day, the moon (and, to a lesser extent, the sun), exerts gravitational pull on the Earth, causing the Earth's oceans to bulge outward. At the same time, the Earth rotates beneath this water, so twice each day, the Earth's coastlines experience high and low tides. These tides, just like rivers and streams, are water in motion. This motion, driven by the pull of gravity, imparts kinetic energy to the oceans. The ebb and flow of the tides along a coast, or perhaps into and out of an inlet or bay, are little different from the flow of water in a river, and they can be harnessed using technology similar to that used on rivers. Because the water flows in two directions, though, the system can generate power when water is flowing in and when it is ebbing out. However, a tidal power-generating station can operate only about ten hours a day, during the times when the tides are in motion.

The oceans' waves are yet another potential source of kinetic energy. Waves, which average about 12 feet (almost 4 meters) in height in the oceans, are caused by wind blowing across the surface of the water, just as tiny ripples are created when a person blows across the surface of a cup of hot chocolate to cool it. The height of a wave—from its peak, or crest, to its bottom, or trough—is determined by how fast the wind is blowing, the length of time it has blown in the same direction, and the width of the open water over which it is blowing. The steepest and most powerful waves are caused by winds that blow strongly in the same direction across oceans, such as the trade winds.

Waves move across the waters of the open ocean with little change. But as they approach the shore and the water gets shallower, they begin to release their enormous energy. First, the ocean's floor causes the wave to slow and to increase in height. Then, the front of the wave "breaks," or collapses, hurling tons of water at the coastline. The force of this wave power is so great that it continues to wash away the coastlines. It is estimated, for example, that parts of Cape Cod are eroding at a rate of 3 feet (0.9 meter) per year. Like the water in rivers and streams, these waves could potentially be used for their kinetic energy.

A final source of kinetic energy in the oceans is their currents. Currents, like waves, are usually propelled by the wind blowing across the surface. The wind has to be strong and consistent. But other currents are formed by differences in water temperature and salinity (salt content) and even by slight differences in the elevation of the sea's surface. The currents follow paths determined by the Coriolis effect, or the effect of the Earth's rotation. In the Northern Hemisphere, the Earth's rotation deflects the currents into a clockwise rotation; in the Southern Hemisphere, the currents flow counterclockwise.

One of the most studied and well-known ocean currents is the Gulf Stream, which originates near Florida, crosses the Atlantic Ocean, and warms much of northern Europe. The Gulf Stream is 50 miles (80 kilometers) wide, and an estimated 10 cubic miles (16 cubic kilometers) of water move through it every hour. It moves so fast that its warm waters do not mix with the colder water that surrounds it. The Gulf Stream is, in effect, a river. The water is in motion, so it contains vast amounts of kinetic energy that could be tapped for human use.

There is also thermal energy, or the heat contained in the world's oceans. Tapping the oceans' thermal energy, though, is not just a matter of somehow going out and piping in the heat. The process, called ocean thermal energy conversion (OTEC), is driven by the ocean's thermal gradient, which refers to the differences in temperature between the ocean's layers of water. Power can be produced when the difference between the warmer surface waters and the colder deep waters is at least 36°F (20°C). Energy-producing systems for tapping the ocean's thermal energy rely on a system of condensers, evaporators, and turbines to generate electricity. OTEC could provide electricity, especially to many tropical nations that currently have to import all their fuel.

Benefits of water energy

The major benefit that all forms of water energy have is that they provide power without burning fossil fuels. Energy can be provided for human use without having to tear up the land to mine coal or disrupt ecosystems to drill for oil. The power they provide is clean—it does not release particulate matter, carbon dioxide, or sulfur dioxide into the air, contributing to smog and the ill health effects that smog can cause, such as lung disease. Also, because water energy does not depend on the burning of fossil fuels, it does not contribute to global warming, caused by the buildup of gases such as carbon dioxide in the atmosphere. Nor does it contribute to acid rain, or precipitation that is more acidic than normal because it contains such substances as sulfur dioxide. Acid rain, like any acidic substance, can have harmful effects on forests, wildlife, and even structures built by people.

Another major benefit of water energy is that it is virtually inexhaustible. Once fossil fuels run out, they are gone. There is no way to somehow manufacture more oil or natural gas. However, the energy provided by water will be there as long as the sun shines and as long as the Earth contains oceans and rivers. Further, the energy provided by water is essentially free—once, of course, the technology is put in place to extract the energy. While money would continue to have to be spent to build plants, maintain them, and distribute the power they produce, a major benefit is that power providers would not have to buy fuel for them. The potential savings is huge. As of mid-2005 the cost of a barrel of oil was hovering around $60. The United States uses about twenty million barrels of oil each day. That means about $1.2 billion per day is spent for just that one form of fuel. Replacing that fuel with water energy would result in enormous savings for consumers.

Drawbacks of water energy

These energy sources, though, are not without their drawbacks. While hydroelectric dams have been around for well over a century, stations for harvesting tidal, wave, ocean current, and ocean thermal power are still in the developmental stages. Exploiting these forms of power would require a huge investment. The cost of building a tidal power-generating station, for example, could run as high as $15 billion.

A second drawback is that water energy is not totally reliable. In an energy plant that burns fossil fuels, the fuel can be fed into the system at a constant rate. As a result, the energy output of the system can be predicted and maintained at a steady pace. Water energy can be a little more variable. In a dry season, the water in a river may not run as fast. The level of the water in the reservoir behind a hydroelectric dam may fall so far that the dam's operators have to slow the flow of water over the dam, cutting power output. In the case of ocean energy, plant operators have no control over the water. Tidal power, for example, can vary from day to day, depending on the alignment of the Earth with the sun and the moon. Wave power could be highly variable, depending on prevailing winds. While the power in ocean currents and in the ocean's thermal gradient is more predictable, the chief obstacle is getting to it. Creating a power plant in the middle of the Gulf Stream would be no easy feat.

A related problem is that water energy is not evenly distributed across the Earth. Providing tidal power to the residents of Nebraska would be impractical because Nebraska is nowhere near an ocean. While tides operate throughout the world, not every coastal region can produce tidal power very efficiently. Some coastal regions have higher tides than others, usually because of some geographical feature, such as bays and inlets that push the water to a higher level than it would otherwise reach. To be practical, efforts to harness tidal power require a difference of about 16 feet (5 meters) between high and low tide. This difference can be found at only about forty places around the world. As the water flows in, and then as it flows out, it can be harnessed in much the same way that the water in any river can be harnessed. However, tidal power stations would be possible only in a limited number of locations.

The use of river power, too, is highly variable. While hydroelectric power provides 24 percent of the electricity used worldwide and 9 to 10 percent used in the United States, much of that hydroelectric power is concentrated in regions with several rivers. In the United States, for example, 14 percent of the power used in the Rocky Mountain states comes from hydroelectric dams; in the Pacific Northwest, in contrast, some 65 percent of power demand is filled by 58 hydroelectric dams. While hydroelectric dams provide almost all of the electricity in Norway, 83 percent in Iceland, 67 percent in Austria, and 60 percent in Canada, they can provide little or none in the desert countries of the Middle East or in most of Africa. This suggests that no one source can magically solve any nation's energy problems.

A final drawback is that a fossil fuel-fired plant can be built essentially anywhere because the fuel is brought to the plant. With water energy, the plant has to be brought to the fuel, meaning that plants have to be built on rivers, along shorelines, and in bays, where they disrupt the natural environment.

Environmental impacts of water energy

A major drawback to the use of water energy is the potential environmental impact. On one level, using water energy would have benefits for the environment, including cleaner air and reduced global warming, compared to the use of fossil fuels. However, the power plants themselves could potentially have a devastating effect on local ecosystems.

Hydroelectric dams are a good example. Throughout the world, about 40,000 large dams are in use to provide hydroelectric power. Most of these dams were built with little regard to the environmental impact they would have. Dams, for example, require reservoirs. In effect, they turn a river ecosystem into a lake ecosystem, at the same time gobbling up large tracts of land. Moreover, they block the migration of fish, such as salmon in the Pacific Northwest. They also prevent the downstream movement of silt, which is often rich in nutrients.

Such facilities as tidal power-generating stations could have similar environmental impacts. The construction and operation of such facilities could have a serious impact on marine and coastal ecosystems, fisheries, and the like. They could disturb the silt on the ocean bed, with unintended consequences. Further, they could convert beautiful natural areas into eyesores.

Another potential drawback to hydroelectric dams—or any water energy project—concerns ownership rights. Rivers usually flow through more than one country. In Southeast Asia, for example, six countries make up the Mekong River's watershed. During rainy seasons this would not be a problem, for the Mekong flows at a rate of 31 cubic miles (50,000 cubic meters) per second. During the dry season, however, the river flows at a rate of only about 1.2 cubic miles (2,000 cubic meters) per second, seriously reducing the amount of power that could be produced. This would provide an upriver country with an incentive to block the flow of the river, denying water and power to the downriver countries. The result could be serious regional conflict over water rights. A similar problem could occur in the oceans. It is an established principle that no country owns the oceans in its vicinity, other than a narrow strip along the coastline. Any type of power-generating station that lies outside of a nation's coastal waters would run into serious legal difficulties if it used international seas to provide power for just one nation.

Economic impact of water energy

The economic impact of water energy has always been great, but new forms have the potential to dwarf the impact that has been felt throughout human history. While water power has been used throughout much of history, its economic impact began to be felt more fully in the late eighteenth and early nineteenth centuries. The town of Lowell, Massachusetts, which grew as textile firms built up around the availability of water power, by the mid-1830s boasted 20 textile mills employing 8,000 people and producing 50 million yards (46 million meters) of cloth per year.

Hydroelectricity had an even larger impact. In the early twenty-first century hydroelectric dams provide about 9 to 10 percent of the electricity used in the United States. Worldwide, though, hydroelectric plants provide about 24 percent of electricity, serving a billion people. Together, they annually produce about 675,000 megawatts (*mega-*, meaning "million"), the equivalent of about 3.6 billion barrels of oil. That represents a savings of about $180 billion that might otherwise be spent on oil. These hydroelectric plants are the world's single largest source of renewable energy.

Other sources of water energy hold even greater promise. Just over 70 percent of the Earth's surface is covered by oceans. The amount of water they contain is staggering: 328 million cubic miles (527 million cubic kilometers), or 361.2 quintillion gallons (1,367.3 quintillion liters). (A quintillion is 1,000,000,000,000,000,000.) Every day the sun shines on these oceans, and every day they absorb a great deal of thermal energy. In fact, the oceans can be thought of as the world's single largest solar panel. It is estimated that on a typical day, about 23 million square miles (60 million square kilometers) of the world's tropical oceans absorb an amount of energy from the sun equal to about 250 billion barrels of oil.

To put that figure in perspective, the total amount of oil produced in the world each day in 2005 was about 76 million barrels. That means that each day, the tropical oceans absorb three thousand times more energy than that provided by oil. This is an enormous amount of energy. Some experts estimate that the amount of power that could potentially be produced from heat in the oceans is 10 trillion watts. Just 1/200th of one percent of this thermal energy—absorbed by the tropical oceans in just one day—could provide all the electricity consumed in the entire United States. This energy would be clean and endlessly renewable. The problem, of course, is finding ways to capture that energy.

Societal impact of water energy

The societal impact of water energy is essentially the same as the impact of any alternative energy. Clean, renewable energy would lessen the adverse health effects of fossil fuel burning. Because the fuel itself is essentially free, more reliance on water power would free up billions of dollars that could be used for other human needs. Using water power would also benefit the environment, reducing the need for environmentally disruptive coal mining and oil drilling, along with the regular oil spills that spoil many nations' coastlines. Water power could also have a major impact on poorer nations, which lack the resources to import fossil fuels for economic development. Water energy could provide these nations with a clean, relatively inexpensive way to develop and provide a richer economic, social, educational, and cultural future for their peoples.

HYDROPOWER

The term *hydropower* is a general one that can be used to refer to any type of water energy. Here, though, the term will be used to refer to the earliest form of hydropower, the kind used in primitive waterwheels, though modern waterwheels are not as primitive as those of the past. In the early 2000s waterwheels continue to be used for low-level electrical power generation.

A waterwheel is a paddlewheel attached to a fixed rotor, or axle, and placed in the current of a river or stream. The wheel is actually a pair of parallel wheels connected to the rotor by radial spokes. Between the two wheels is an arrangement of paddles. As the water passes, the kinetic energy of the water pushes against the paddles, turning the wheel and producing mechanical energy, which in turn is transferred through gears to machinery that accomplishes the task at hand. In the past this machinery was very often a large stone used to grind grain, but could also consist of saws in a sawmill, bellows in a foundry, looms in a textile mill, abrasive tools for polishing metal, pumps for removing water from a mine, and many other applications. Some wheels, rather than using paddles, used buckets. The weight of the water in the buckets helped to propel the wheel around.

Early waterwheel users were creative with the placement of waterwheels. While the wheels were often inserted directly into a stream or river and connected to a facility on the riverbank, often they were placed on barges and boats (called ship mills), sometimes suspended between two barges or boats. Others were attached to the abutments of stone bridges over rivers.

The John Cable Mill in Cades Cove, Tennessee, was in operation up to the mid-1900s to grind corn and saw logs. Two streams were used to provide adequate water flow to the waterwheel in order to generate power. *James Steinberg/Photo Researchers, Inc.*

Historically, three different types of waterwheels were used. The first was the horizontal waterwheel. This type of wheel was lowered horizontally into the water, where it was totally submerged. Attached to the wheel were veins, which were somewhat like the veins on a pinwheel that turns when air blows over it. This type of wheel was attached to a rotor that protruded up out of the water and connected directly to something like a millstone. Horizontal waterwheels are still in use in India and Nepal.

A more efficient and powerful design is the vertical waterwheel. Vertical waterwheels came in two types, the undershot and the overshot, both of which required a system of gears to turn the machinery. An undershot wheel was lowered vertically into the water of a river. The water passed by the lower portion of the wheel, pushing on the paddles to turn it. A major disadvantage of this type of wheel was the variability in the river's water level. During dry spells, the water level in the river would fall, diminishing the wheel's power. Sometimes the water level would fall so much that the wheel was entirely out of the water, making it useless.

With an overshot wheel, the water flowed from above. These types of wheels were sometimes positioned underneath waterfalls so that the water struck the paddles as it fell, or alternatively poured into buckets so that the weight propelled the bucket forward, turning the wheel. More commonly, the source of the water was an artificial channel that flowed to a position above the waterwheel.

Current uses of hydropower

Although waterwheels are thought of as a feature of earlier societies, in fact they are still widely used for irrigation, pumping water, and even occasionally still to power machinery such as sawmills. These types of wheels can be found in many areas of the world. In Turkey and Afghanistan, waterwheels are still used to grind grain. In the United States, a company called Equality Mills in West Virginia still manufactures waterwheels, and one of the first wheels the company ever produced, in 1852 (under earlier owners), is still in operation at the Tuscorora Iron Works just across the creek.

Companies in the United States and Germany also manufacture waterwheels for electrical power generation, and the British Hydropower Association provides detailed information about building small waterwheel power plants. Typically, such a plant would involve the following:

- A water intake from a river or stream
- A small canal to channel the water
- A forebay tank, where the water is slowed so that debris can settle out, along with a trash rack to filter out debris
- A penstock, which shoots the water downward to the turbine
- A powerhouse, which contains a turbine where the power is actually generated
- A tailrace, which channels the water back into the river or stream

Benefits of hydropower

Prior to the industrial revolution, waterwheels were essentially the only form of alternative energy available. In Europe, the rapid spread of waterwheels may have been a function of the Black Death, the plague that wiped out large portions of the population in the late Middle Ages. Waterwheel use expanded rapidly in England, France, and other European nations as a way to replace lost labor.

In modern times waterwheels are used primarily for low-level electrical power generation. The British Hydropower Association notes that small-scale hydropower generation is highly efficient, between 70 and 90 percent (meaning that 70 to 90 percent of the available power can actually be generated).

Drawbacks of hydropower

Historically, waterwheels had two primary drawbacks. The first was that they required a great deal of maintenance. Because they were constructed mostly of wood, they tended to break down over time. Further, water is not very friendly to wood, causing it to deteriorate and rot. The second problem was that in northern climates, waterwheels were of limited usefulness in cold weather, when the water froze.

The primary drawback of modern waterwheels is that building such a power plant is expensive for the amount of energy it can produce. The bulk of the expense lies in the turbines needed to generate the power, gearboxes needed to convert kinetic energy into mechanical energy, and generators needed to convert mechanical energy into electrical energy. The extent to which this is a drawback depends on the amount of available energy. When flow is high, the amount of power generated is more likely to justify the cost; when it is low, the amount of power generated may not be worth the cost. The British Hydropower Association estimates that the total cost of building a 100-kilowatt (kW) power plant could range from roughly $150,000 to $470,000. Adding to the cost is the need to acquire rights to use the land.

Another potential drawback of waterwheel power plants is safety. Such plants, including the wheel itself, have to be fenced off so that they do not injure curious people who get too close. This fencing, combined with the plant itself, has the potential to become an eyesore, though manufacturers attempt to make the equipment as visually attractive as possible.

A final drawback stems from the variability of water flow. During spring runoff, when snow is melting and rivers run rapidly, the amount of power generated is much higher than in, say, August, when rivers are running low, providing less flow.

Issues, challenges, and obstacles of hydropower

The primary issue surrounding the use of waterwheels is ownership rights. Any stream or river almost certainly flows through property owned by many people. The river itself is common property; no one individual owns it. If one property owner builds a waterwheel, other property owners along the river might object, particularly if they are uncertain about the effects the wheel might have downstream.

Another challenge concerns distribution of the power. One property owner might build a waterwheel for personal use, but larger waterwheels in high flow streams might generate enough electricity for multiple users. The questions then become how that power is going to be distributed and how its users will divide the cost of constructing the waterwheel.

HYDROELECTRICITY

Hydroelectricity is any electricity generated by the energy contained in water, but most often the word is used to refer to the electricity generated by hydroelectric dams. These dams harness the kinetic energy contained in the moving water of a river and convert it to mechanical energy by means of a turbine. In turn, the turbine converts the energy into electrical energy that can be distributed to thousands, even millions, of users.

One of the most prominent hydroelectric dams in the United States is the Hoover Dam on the Colorado River along the border between Arizona and Nevada. Construction on the dam began in 1931; it was completed five years later, under budget, for $165 million. Behind the dam is a reservoir, Lake Mead, containing about 1.24 trillion gallons of water. The dam is 726 feet (221 meters) tall, and at its base is 660 feet (201 meters) thick. Its 4.5 million cubic yards of concrete would be enough to build a two-lane highway from Seattle, Washington, to Miami, Florida. Each year, the dam produces 4 billion kilowatt-hours of electricity, enough to serve 1.3 million people.

The largest hydroelectric dam in the United States is the Grand Coulee Dam on the Columbia River in Washington State. Construction began on the dam in 1933 and was completed in 1942. The original purpose of the dam, however, was not to generate electricity but to irrigate one-half million acres of agricultural land. From 1966 to 1974 the power-producing ability of the dam was expanded with the addition of six new electrical generators. The scope of the Grand Coulee Dam amazes visitors. It is the largest concrete structure in the United States, at 11,975,521 cubic yards. At its widest point, it is almost exactly a mile (1.6 kilometers) long. At 550 feet (167 meters) tall, it is twice the height of the Statue of Liberty and more than twice the height of Niagara Falls. Its reservoir, Roosevelt Lake, contains up to 421 billion cubic feet of water. Its four power plants and 33 generators produce 6,809 megawatts of power annually.

Huge turbine engines inside the Hoover Dam in Black Canyon, Nevada, supply electricity and water to California, Nevada, and Arizona. © *James Leynse/ Corbis.*

A hydroelectric dam consists of the following components:

- Dam: The dam is built to hold back water, which is contained in a reservoir. This water is regarded as stored energy, which is then released as kinetic energy when the dam operators allow water to flow. Sometimes these reservoirs, such as Lake Mead, are used as recreational lakes.

- Intake: Gates open to allow the water in the reservoir to flow into a penstock, which is a pipeline that leads to the turbine.

Aerial view of Hoover Dam, Nevada, which was built between 1931 and 1936 to harness the Colorado River, creating the reservoir Lake Mead. © *Lester Lefkowitz/ Corbis.*

The water gathers kinetic energy as it flows downward through the penstock, which serves to "shoot" the water at the turbine.

- Turbine: A turbine is in many ways like the blades of a windmill or the veins of a pinwheel. The water flows past the turbine, striking its blades and turning it. The most common turbine design used in large, modern hydroelectric power plants is the Francis turbine, which is a disc with curved blades. The Francis turbine was developed by British-American engineer James B. Francis (1815–1892), who began and ended his professional career in the United States as an engineer at the Locks and Canal Company in Lowell, Massachusetts. In the largest hydroelectric plants, these turbines are enormous, weighing up to 170 tons or more. The largest ones turn at a rate of about 90 revolutions per minute.

- Generator: The turbine is attached by a shaft to the generator, which actually produces the electricity. Generators are based on the principle of electromagnetic induction, discovered by

Roll on, Columbia

In the 1940s folk singer Woodie Guthrie (1912–1967) was hired by the Bonneville Power Administration to write folk songs about the dams being built on the Columbia River. Over a period of about a month, Guthrie wrote twenty-six folk songs under the general title *Columbia River Ballads.* One of the most popular of these songs was "Roll on, Columbia," which the state of Washington adopted as its official folk song in 1987.

British scientist Michael Faraday (1791–1867) in 1831. Faraday discovered that as a metal that conducts electricity, such as copper wire, moves through a magnetic field, an electrical current can be induced, or created, in the wire from the flow of electrons. The mechanical energy of the moving wire is therefore converted into electrical energy. In a hydroelectric plant, the mechanical energy is supplied by the turbine, which in turn is powered by the kinetic energy of moving water.

- Transformer: A transformer converts the alternating current produced by the generator and converts it into a higher voltage current.

- Power lines: Power lines transmit the power out of the power plant to the electrical grid, where it can be used by consumers.

- Outflow: Pipes called tailraces channel the water back into the river downstream.

Hydroelectric power plants come in three basic types:

- High head: "Head" refers to the difference in level between the source of the water and the point at which energy is extracted from it. Assuming other things are equal, the higher the head, the more power is generated. A high head hydroelectric plant is one that uses a dam and a reservoir to provide the kinetic energy that powers the plant. Most major hydroelectric plants are of this type.

- Run-of-the-river: In contrast, a run-of-the-river plant requires either no dam or a very low dam. It operates entirely, or almost entirely, from the flow of the river's current. No energy is stored in a reservoir. These hydroelectric plants are generally small, producing less than about 25 kilowatts.

The World's Biggest Hydroelectric Power Plant

The world's biggest hydroelectric power plant is in South America. From 1975 to 1991 the Itaipú Dam was built across the Paraná River as a joint project by Brazil and Paraguay. The plant has eighteen generating units that can provide 12,600 megawatts of power, or 75 million megawatt-hours per year, enough wattage to power most of California. By 1995 the dam was providing 25 percent of Brazil's energy and 78 percent of Paraguay's.

The dam, called one of the "Seven Wonders of the Modern World" by the American Society of Civil Engineers, is enormous. The amount of iron and steel used in its construction could have built 380 Eiffel Towers (the famous landmark in Paris). The volume of concrete used to construct it is equal to fifteen times the volume used to construct the tunnel under the English Channel that connects France and England. To build the dam, workers had to rechannel the seventh largest river in the world and remove 50 million tons of earth and rock.

- Pumped-storage: Some hydroelectric plants rely on a system of two reservoirs. The upper reservoir operates exactly as the reservoir does in a high head plant: Water from the reservoir flows through the plant to turn the turbines, then exits the plant and reenters the river downstream. In a pumped-storage plant, the water exiting the plant is stored in a lower reservoir rather than reentering the river. Using a reversible turbine, normally during off-peak hours (or hours when power usage is low, usually at night), water is then pumped from the lower to the higher reservoir to refill it. This gives the plant more water to use to generate electricity.

Current uses of hydroelectricity

During the 1930s a large number of hydroelectric dams were built on the waterways of the United States. Many of these dam projects were the result of that decade's Great Depression. During the depression, the U.S. government sponsored public-works projects designed to put people to work and recharge the economy. These dams, such as Hoover Dam and the 192 dams that were built along the Columbia River in the Northwest, produced

The World's Smallest Hydroelectric Power Plant

Someday soon the world's smallest hydroelectric power plants may appear—in people's shoes. On file at the U.S. Patent and Trademark Office is patent number 6,239,501. The patent is held by Canadian inventor Robert Komarechka, who conceived the idea that a tiny hydroelectric power plant embedded in the soles of shoes could provide power to run cell phones, compact-disc players, laptop computers, and other modern electronic gadgets.

The design is based on the way people walk. When a person takes a step, force is exerted downward on the heel. The foot then rolls forward, so that force is exerted on the toe. Komarechka found a way to harness this power by inserting sacs of fluid in the soles of shoes, one at the heel end and one at the toe end. Connecting the sacs is a conduit through which the fluid, a gel-like substance, can flow. As it flows, it turns a tiny turbine that is attached to a microgenerator, which in turn produces electrical power. A tiny socket allows the user to connect an electronic gadget to the power source, either directly at the shoe or at a power pack attached to, perhaps, a person's belt.

hydroelectric power, and by the end of the 1930s they were meeting about 40 percent of the nation's electricity needs.

Many dams were also built in a seven-state region around the Tennessee River Valley under the guidance of the Tennessee Valley Authority (TVA). In the early twenty-first century about two thousand hydroelectric dams in the United States provide about 9 to 10 percent of the nation's electricity. They have not kept pace with U.S. demand for power simply because most of the best sites for hydroelectric dams already have one. Worldwide, about 40,000 hydroelectric dams provide a total of 675,000 megawatts of power to a billion users.

Benefits of hydroelectricity

The chief benefit of hydroelectric power, like the power provided by waterwheels, is that fossil fuels do not have to be burned, releasing particulate matter and greenhouse gases (such as carbon dioxide and sulfur dioxide) into the atmosphere, where they produce smog and contribute to global warming and acid rain.

Hydroelectric power is also free in the sense that fuel does not have to be purchased to produce it, although of course money has to be spent to build and maintain the power plant and to distribute power to consumers.

Another major benefit of hydroelectric energy is that it is renewable. Over time, it will become more and more expensive to extract fossil fuels from the earth until eventually these fuels will be entirely depleted. Hydroelectric power will remain available as long as there are rivers. Hydroelectric energy, in contrast to oil, is not dependent on imported fuels from other countries, which could be cut off by one or more of those countries and make a nation vulnerable to political pressures from them. Hydroelectric dams can also have secondary benefits. They provide flood control on rivers, and their reservoirs often serve as lakes for recreational activities such as boating and swimming.

Drawbacks of hydroelectricity

Hydroelectric energy has always been thought of as clean energy, but scientists and engineers have started to understand

Chinese workers inspect the second section of the main dam of the Three Gorges project near Yichang in central China's Hubei province. China blocked the massive Yangtze River on June 1, 2003, to fill a reservoir for the world's biggest hydroelectric project that is a point of national pride but which critics fear will become an environmental nightmare. *©Reuters/Corbis.*

that it has significant drawbacks as well. One drawback is that damming rivers floods large areas of land. When the water fully rises behind the new Three Gorges Dam on China's Yangtze River (under construction in 2005), for example, it will wipe out 13 cities, 140 small towns, and over 1,300 small villages, forcing over two million people to leave some of China's richest farmland. In Quebec, Canada, the first phase of a major hydroelectric project on the watershed flowing into the James Bay flooded nearly 3,900 square miles (10,000 square kilometers); the second phase of the project more than doubled that figure. A third phase of the project was still in the planning stages in 2005, but if the entire project is carried out as planned, the size of the flooded regions would be greater than the size of the country of Switzerland. Flooding vast amounts of land like this often has a disproportionate effect on native peoples, whose way of life can be destroyed.

Constructing hydroelectric dams, converting a free-flowing river of fresh water into a lake, also has a profound effect on ecosystems. Dams and reservoirs affect such factors as water quality, the amount and kinds of bacteria in the water, bank erosion, nutrient transport, the salt content of soil, and water temperature. Some dams have been implicated in the spread of waterborne diseases such as malaria. When a large dam fails, the results can be catastrophic, wiping out wildlife, vegetation, houses, roads, even whole towns downstream.

Dams also affect the amount of water in rivers downstream, with effects on wildlife that are only beginning to be understood. They also block the flow of silt downstream, affecting the flow of nutrients through a river system. In Egypt, the Aswan Dam along the Nile River, which provides 10 billion kilowatt-hours of electricity every year (and has a reservoir of nearly 6 trillion cubic feet [170 million cubic meters], four times that of the Hoover Dam), blocked the flow of nutrient-rich silt to the nation's agricultural floodplains. Farmers have had to replace those nutrients with a million tons of artificial fertilizer each year. Meanwhile, the silt can build up at the dams over time, causing them to be less efficient.

Some scientists estimate that 93 percent of the declines in freshwater marine life are caused by hydroelectric dams. The dams in the U.S. Pacific Northwest are regarded as a major cause in the decline of the salmon population because the dams prevent salmon from migrating upriver to spawn. Although "fish ladders" are installed to lessen this impact, they are by no means 100 percent effective.

Another drawback is that hydroelectric energy may not be as clean as once thought. Decaying vegetation in reservoirs may give off quantities of greenhouse gases equal to those emitted by burning fossil fuels. This can be an ongoing problem because when the water level in a reservoir falls during an extended dry period, vegetation grows on the banks. This vegetation, then, is covered by water when the reservoir refills during wet periods, causing the vegetation to rot again and emit gases such as methane and carbon dioxide, contributing to global warming. Finally, this decaying vegetation can alter the form of mercury contained in rocks into a form that is soluble in water. Mercury, a heavy metal like lead, can accumulate in the tissues of fish. It thus poses a health hazard to people who consume the fish.

A general view taken June 7, 2003, of old Wushan county, near China's Chongqing Municipality, which was partially submerged by rising water levels after China blocked the massive Yangtze River for the Three River Gorges project. ©*Reuters/ Corbis.*

Economic impact of hydroelectricity

As of 2005 there are about 40,000 large hydroelectric dams in operation worldwide (a large dam is defined as one that is taller than a four-story building, or more than 49 feet [15 meters]). The country with the greatest number of large dams is China, with 19,000. The United States is second, with 5,500. Major dams are defined as those more than 492 feet (150 meters) in height. The United States leads the world with 50 major dams.

The economic impact of hydroelectric power can be considerable. In some countries, such as Norway, hydroelectric dams provide virtually all of the nation's electrical needs. In Canada, about 60 percent of the nation's electricity is provided by hydroelectricity. Canada, and especially the province of Quebec, provides a good example of the economic impact of hydroelectricity. In the 1960s Quebec launched a program to foster economic development. One of the centerpieces of this program was the development of hydroelectric power in the James Bay region of northwestern Quebec. The first phase of the project began in 1972, when three rivers—the Caniapiscau, Eastmain, and Opinaca—were diverted into reservoirs. These reservoirs, along with a system of 215 dikes and dams and four power stations, nearly doubled Quebec's hydro-power production. Construction employed 12,000 people and required 203 million cubic yards (0.9 million cubic meters) of fill dirt and rock, 138,000 tons of steel, 550,000 tons of cement, and 70,000 tons of explosives—all of which provided economic opportunities for Canadians. This first phase of the project, completed in 1985, provided 10,300 megawatts of electricity at a total cost of $14 billion.

Construction on the second phase of the project began in 1989, but it was suspended in 1994, when the project was nearly complete, because of environmental concerns, as well as objections raised by the Cree, a native community that lived in the James Bay region. These problems were resolved, and construction was completed in 2002. Combined, the two phases of the project produce 15,000 megawatts of electricity, or three times the amount of power produced by Niagara Falls. A third phase of the project was scheduled to begin in 1989, but that phase was put on indefinite hold because of environmental concerns. In large part because of the James Bay project, Quebec's electrical output increased from 3,000 megawatts in the early 1960s to 33,000 megawatts in 2002. Further, in 1997 Canada sold about $600 million in electrical power to the United States; by 2002 that figure had climbed to $3.5 billion. Ninety-three percent of this electricity is hydroelectric power.

Societal impact of hydroelectricity

The negative societal impact of hydroelectric power development is often felt most by native peoples. In northern Quebec, the

Cree, an Algonquin-speaking people, were profoundly affected by the James Bay project. In 1975 the Cree were awarded $225 million in compensation for the disruption that the project caused in the Cree way of life, which revolved around fishing, hunting, and fur trapping in the watershed around James Bay. That money, however, could not compensate the Cree for the immense changes the project caused in Cree society. One Cree band, or tribe, was forced entirely off the land. Among the two remaining bands, the hydroelectric project (along with other enterprises such as mining and lumber) virtually destroyed hunting and trapping grounds, threatening the economic and cultural survival of the Cree.

This type of social problem is not limited to Quebec. In the United States, the construction of the Grand Coulee Dam in Washington State forced the Colville Indian tribe off their traditional hunting and fishing grounds. The Colville tribe sued the federal government and in the 1990s was awarded a $52 million lump-sum settlement. An organization called the International Rivers Network estimates that worldwide, between thirty and sixty million people, about two million a year, have been displaced (driven off their land) by hydroelectric dams. In most cases, the displaced people are small farmers and native peoples.

Issues, challenges, and obstacles of hydroelectricity

Hydroelectric power faces many obstacles. It is estimated that the amount of hydroelectricity available is about four times the amount being used. The United States has over 5,000 sites that have been identified as possible sites for hydroelectric dams. Many other sites have been identified in Asia and Africa. However, hydroelectric projects often meet with much resistance from environmental groups and others who are concerned about the effects of hydroelectric dams. In the past, the World Bank was willing to loan money to countries to build dams. In more recent years, largely because of environmental concerns and the effect of dams on native peoples, the World Bank has provided less money for these projects.

Research continues on the impact such dams have on fish populations, along with ways to minimize this impact. Research also continues on ways to improve water quality and dam safety, as well as ways to improve the efficiency of hydroelectric dams. In the United States, numerous efforts have been made to "uprate," or improve the efficiency, of older dams. The result since the late

1970s has been to add about 1.6 million kilowatts to the nation's power supply without building new dams. This power costs less than one-fifth of the cost of electricity produced by new oil-fired generators.

OCEAN THERMAL ENERGY CONVERSION

Ocean thermal energy conversion, or OTEC, is the primary means of extracting thermal energy from the world's oceans. It is based on the thermal gradient, which refers to the difference in temperature between the ocean's surface waters, which are warmed by the sun, and its deeper waters, which originate in polar latitudes and are therefore much colder. The concept of using the thermal gradient to produce electricity was first proposed by French biophysicist Jacques Arsene d'Arsonval (1851–1940) in 1881. D'Arsonval proposed the basic form of a system that is still used.

OTEC is based on two different technologies, closed cycle and open cycle, which can be combined into a hybrid system as well:

- Closed cycle: The system that d'Arsonval envisioned was a closed-cycle system. The working fluid was ammonia, which boils at a low temperature, 28°F (33°C). Heat transferred from the warm surface waters of the ocean boils the ammonia. As the vapors expand, they turn a turbine, which is connected to a generator that produces electricity. Cold seawater pumped up from depths of 2,625 to 3,280 feet (800 to 1,000 meters) is used to condense the ammonia vapor in a condenser. The ammonia is then recycled back through the system.

- Open cycle: In an open-cycle system, the working fluid is the warm surface water itself. In a near vacuum, the warm water vaporizes at the surface-water temperature. Like the ammonia vapor in the closed-cycle system, the expanding water vapor drives a turbine, which is attached to a generator that produces electricity. The open-cycle system has the added advantage of producing desalinized water, or water from which the ocean's salt has been removed. Thus, when the water is condensed by the cold water pumped from the depths, it can be siphoned off and used as drinking water. The underlying process is little different from the condensation that forms on a glass of iced tea on a humid summer day. Unlike the closed-cycle system, in which the ammonia is recycled again and again, the open-cycle system operates with a continuous supply of warm seawater.

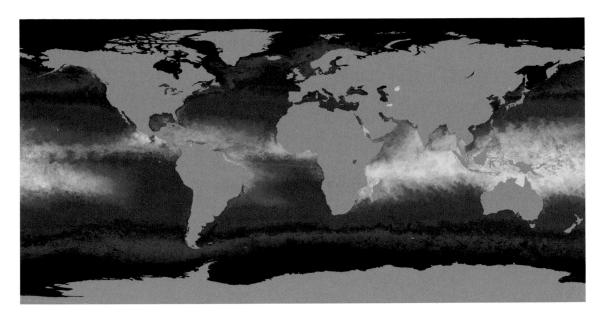

- Hybrid systems: Hybrid systems employ both closed- and open-cycle systems, getting the benefits of each. The closed-cycle system produces more electricity than the open-cycle system, but the open-cycle system produces fresh water as well as electricity.

Current uses of ocean thermal energy conversion

Most research on OTEC is conducted by the Natural Energy Laboratory of Hawaii Authority (NELHA), formed in 1974. NELHA conducted the first at-sea test of a closed-cycle plant in 1979. The project was called Mini-OTEC, and it took place on a converted navy barge off the coast of Keahole Point, Hawaii. For three months the plant generated 50 kilowatts of gross power. The plant pumped 2,700 gallons (10,220 liters) per minute of cold (42°F [5.5°C]) seawater up from a depth of 2,200 feet (670 meters). The plant pumped an equal amount of warm (79°F [26°C]) surface water. Some of the plant's power had to be used to run the pumps, so the net power output of the plant ranged from 10 to 15 kilowatts.

From 1992 to 1998 NELHA conducted a major demonstration project at its Keahole Point facility. It designed and built a 210-kilowatt open-cycle plant. At its peak the plant produced about 255 kilowatts of power. However, it generally used about 200 kilowatts to pump 6,500 gallons (24,605 liters) per minute of

Colored computer model of global sea temperatures in 2001, based on satellite data. The surface temperature of the Earth's oceans has been color-coded and combined with a projection of the land surface (gray). The temperature varies from a warm 35 degrees Celsius (yellow) in the tropics, through red, blue, purple, and green to a freezing minus 2 degrees Celsius (black) in the polar regions. *NASA/Photo Researchers, Inc.*

43°F (6°C) water from a depth of 2,700 feet (823 meters) and 9,600 gallons (36,340 liters) per minute of 76° to 81°F (24° to 27°C) surface water, for net power of some 50 to 55 kilowatts. Its highest net power output was 103 kilowatts, along with production of about six gallons (22 liters) per minute of desalinated fresh water. Designs were drawn for a 1.4-megawatt plant with the potential to produce about 400 net kilowatts, but funding was unavailable, so the project put on hold. As of the early 2000s no OTEC plant is operating anywhere in the world.

Benefits of ocean thermal energy conversion

OTEC draws on natural resources that are renewable, abundant, and clean. Rather than burning fossil fuels, OTEC power plants rely on warm seawater on the oceans' surfaces and cold seawater from their depths. By replacing such fuels as coal and oil, they can help eliminate the need for mines and oil-drilling platforms, which are not only unsightly but also are potential sources of pollution. Further, the amount of solar energy absorbed by the oceans, particularly in tropical climates, is far in excess of current human energy needs. Unlike wind and tidal energy, thermal energy is always present at consistent levels, which would make it an extremely reliable source of energy.

A second benefit is that OTEC plants do not release greenhouse gases such as carbon dioxide that contribute to global warming, nor do they release sulfur dioxide, a chief cause of acid rain. Further, scientists have concluded that discharging water back into the oceans has only minimal environmental drawbacks. A third benefit is that OTEC can reduce dependence on imported fuel. A state such as Hawaii, as well as many nations around the world, has to import most or all of its fuel. This need to import fuel both drains cash from the economy and makes the state or country dependent on other countries for its energy needs.

Finally, OTEC has a number of secondary benefits. It produces fresh water as well as electricity, a potentially major benefit for countries in which the amount of fresh water is limited. The amount of fresh water created can be up to 1.3 gallons for every 264 gallons (5 liters for every 1,000 liters) of cold seawater in an open-cycle plant. The cold seawater in OTEC can also be used to air-condition buildings, and contribute to mariculture, the cultivation of fish, shellfish, kelp, and other plants that grow abundantly in cold water. Also, eighty-four of the Earth's elements are in solution in the oceans' waters in trace amounts. Some of these elements, such as magnesium and bromine, have commercial value and could be efficiently extracted from the water used in OTEC.

Drawbacks of ocean thermal energy conversion

The major drawbacks to OTEC are geographical and economic. OTEC plants have to be located in places where the difference in temperature between the warm surface waters and cold deep-sea waters is great enough—at least 36°F (2°C); 40°F (4°C) would make the plant even more efficient. For shore-based plants, this difference would have to be present fairly close to the shore, although floating OTEC ships could expand the range of plants' geographic locations.

OTEC faces a number of economic obstacles. The cost of producing electricity through OTEC is higher than the cost of producing it from fossil fuels. Presently, there is not enough economic incentive for nations to invest billions of dollars in OTEC plants. Scientists and engineers estimate that after the high initial construction costs, the electricity produced over a long period, perhaps thirty years, would be economical, but no one knows how long these types of plants could function without requiring a major overhaul. Scientists and engineers are continuing to work on the development of major OTEC components to make them more durable, more efficient, and less costly.

Environmental impact of ocean thermal energy conversion

OTEC has very little in the way of environmental impact. The only hazardous substance is the working fluid, which in the case of closed-cycle plants is ammonia. However, the ammonia is recycled through the system, so an OTEC plant does not release any noxious substances into the water or atmosphere. An open-cycle plant releases some carbon dioxide, but the amount is 1 percent of the amount released by fuel-oil plants per kilowatt-hour.

What needs to be tested in a large commercial or experimental station is the effect of an OTEC plant on water temperatures and on marine life in the upper layer of the water. An OTEC plant pumps cold, nutrient-rich water from the depths up to the surface. This mixing of different temperatures of water could have effects on marine life that are currently not well understood. OTEC engineers are also concerned about the potential effects on fish populations. The discharge of nutrient-rich water could increase fish populations in the vicinity of a plant. On the other hand, the plant itself could also disrupt spawning patterns or result in the loss of fish eggs and tiny young fish. Again, these potential environmental impacts are not known.

Economic impact of ocean thermal energy conversion

Given current technology and the cost of fossil fuels, the economic impact of OTEC would most likely be greatest for small island nations that have to import all their fuel. Such a country, for example, Nauru in the South Pacific, would be able to benefit from a 1-megawatt plant. Such a plant could produce electricity for pennies per kilowatt-hour. It has been estimated that a 100-megawatt OTEC plant could produce electricity for about $0.07 per kilowatt-hour. The chief problem, however, is the initial cost of construction. That same 100-megawatt plant would cost about $4,200 per kilowatt capacity, or about $420 million. It is unlikely with the cost of fossil fuels relatively low that nations will make this type of investment. However, as of 2005 the cost of fuel oil was rising and reached $60 per barrel. If fuel oil continues to become more expensive, OTEC may become more of an option, and organizations such as the World Bank may become more willing to loan funds for construction.

Issues, challenges, and obstacles of ocean thermal energy conversion

The chief obstacle to OTEC development is the high initial construction cost of such a plant. Researchers continue to find ways to bring down the construction costs, particularly to reduce the cost of condensers and other components of the system. Research is also being conducted to find ways to boost the net power output of the system—that is, the amount of power left over after a portion of the power is used to pump water through the system. As of 2005, governments and international organizations remained reluctant to provide funds for the development of OTEC plants, whose long-run benefits are not entirely clear.

TIDAL POWER

Tidal power refers to the use of the oceans' tides to generate electricity. Sir Isaac Newton (1642–1727) pointed out in the seventeenth century that every day, the gravity of the moon exerts a pull on the Earth. This gravitational pull has little effect on the Earth's solid landmasses. But the oceans' waters are fluid, so as the moon's gravity pulls on them, they bulge outward. These bulges, which place along an axis (an imaginary line) that points toward the moon, are called lunar tides; on the other side of the Earth, the side away from the moon, the waters bulge out away from the gravitational pull of the center of the Earth.

Annapolis Royal Tidal Generator, a hydroelectric power station in Nova Scotia, is located in Annapolis Royal by the Bay of Fundy, home of the world's highest tides. Twice a day, the tide comes in and out. Twice a day the turbine turns. Twice a day electricity is generated and supplied to the provincial electric grid. *Stephen J. Krasemann/Photo Researchers, Inc.*

While the moon does most of this work, the sun helps out, but to a lesser extent. This is because the gravitational attraction one body has on another is the result of two factors: its size and its distance. Although the sun is much bigger than the moon, the moon is much closer to the Earth, so it exerts a greater gravitational pull. Nonetheless, the sun's gravitational pull also creates tides, called solar tides.

When the Earth, moon, and sun are aligned in a straight line during a full or new moon, both the sun and moon are pulling together in the same direction, like two teammates in a tug-of-war. During a full moon the pull is greatest, creating large tides called spring tides. During half-moon periods, when the moon and sun are at right angles, or 90 degrees, to each other, the tides created, called neap tides, are lower, simply because the lunar tides are being pulled out along one axis and the solar tides along a perpendicular axis. During these times the coasts have two low and two high tides over a period of less than twenty-four hours.

At the same time, the Earth rotates beneath these bulges, passing under each one during a twenty-four-hour period. The result is

Tidal Power Forever?

In most discussions of tidal power, one of the chief advantages cited is that tidal power is endlessly renewable—that the Earth will never run out of it because the tides will always be there. Technically, this claim is not entirely true. The bulging oceans exert friction on the Earth, gradually slowing down the speed of the Earth's rotation. This means that in time, tidal power will no longer exist.

As a practical matter, though, this is no cause for concern. This slowing of the Earth's rotation will not have any significant effect for billions of years! By that time humankind will no doubt have harnessed a form of power that cannot be imagined today. In the meantime, scientists have calculated that harnessing all of the tidal power of the oceans would slow the Earth's rotation by twenty-four hours every two thousand years.

that tides rise and fall rhythmically along the world's coastlines approximately twice each day in predictable patterns. These flows of water are very like the flows of rivers, and their energy can be harnessed in much the same way that a river's energy is by a hydroelectric dam.

There are two ways to harness energy in tidal power-generating stations: the tidal barrage and tidal streams. A tidal barrage, also called an ebb generating system, is very similar to a dam. The barrage is constructed at the mouth of a bay or estuary (a water passage where the tide meets the lower end of a river). For a barrage to be workable, the difference in water elevation between low tide and high tide has to be at least 16 feet (5 meters).

When the tide flows in, the water moves through moveable gates in the barrage called sluice gates, similar to a "doggy door" a family pet can use to enter the house just by pushing on it. When the tide stops flowing in, the gates are closed, trapping the water in a basin. The water now represents stored energy, in much the same way that the reservoir behind a hydroelectric dam does. As the tide then flows out (ebb tide), the gates in the barrage are opened. This allows the water to turn turbines as it flows back out to sea. Just as in hydroelectric plants, the turbines are connected to a generator, which produces electricity. It is possible to have flood-generating systems, where the water turns the turbines as it flows in rather than out, but hydrologists and engineers believe that these systems are less efficient. It is also possible to have systems that work in both directions, but these kinds of systems would be difficult and more expensive to build because the turbines that would have to be used would have to work in both directions. Consequently, the best design for most sites is the ebb-generating system.

Other technologies exist for harnessing tidal power, but all these technologies are in early stages of development. In each case, the goal is to tap the energy contained in tidal streams. A tidal stream is a fast-flowing current of water caused by the movement of the tides. These streams can occur wherever a natural barrier constricts the flow of water, which then speeds up after it passes the constriction. Thus, a tidal stream might flow between two islands, or between the mainland and an offshore island. The chief advantage of these technologies is that a tidal basin does not have to be constructed.

Current use of tidal power

Currently, only one major tidal power generating station is in operation. This station is located on the estuary of the La Rance River in France. Construction of the barrage began in 1960 and was completed in 1966. The barrage is almost 1,100 feet (330 meters) long with a 13.7-square-mile (22-square-kilometer) basin. The station uses twenty-four turbines, each 17.7 feet (5.4 meters) in diameter. Each turbine is rated to produce about 10 megawatts of power, so the station can produce a maximum of 240 megawatts. (To put that figure in perspective, the average coal- or oil-fired power plant produces about 1,000 megawatts.) There are 8,760 hours in a year, so the system can produce 2,102,400,000 kilowatt-hours per year, enough to supply most of the electricity needs of the Brittany region of France.

Other nations have explored the possibility of harnessing tidal power. Since the 1960s tidal power has been proposed in the Kimberley region of western Australia. There, it was estimated that tidal power could provide 3,000 megawatts of electricity. Australia's Renewable Energy Commercialisation Program awarded a grant to develop a 50-megawatt plant in the Derby region of Australia. Scotland, too, has explored tidal energy, and proposals have been made for the construction of a tidal station on Solway Firth in southwest Scotland; in the 1970s Scotland built a 15-kilowatt experimental tidal turbine on Loch Linnhe. In England, the Severn River has been identified as a promising site for a tidal power station. The most promising site in the world is the Bay of Fundy in Canada, which, at up to 56 feet (17 meters), has the highest tides in the world.

Benefits and drawbacks of tidal power

The chief benefits of tidal power, as of most forms of alternative energy, are that it is clean, renewable, and does not consume resources such as coal or oil. It does not discharge pollutants into the water or atmosphere, so it does not contribute to acid rain or global warming. Further, the energy source is free. Tidal power barrages have a secondary benefit, for they can function as bridges linking communities on opposite sides of an estuary, making travel quicker.

The chief drawback of tidal power stations is their expense. It has been estimated, for example, that construction of a tidal power station on the Severn River in England would cost about $15 billion. A second drawback is that not every coastal region is suitable for tidal power. Generally, a difference between high and low tides of about 16 feet (5 meters) is necessary for a tidal power station to be cost-effective. Only about forty such sites in the world have been identified. A third drawback is that the tides are in motion only about ten hours per day. This means that tidal power cannot be provided consistently throughout the day and would have to be supplemented with other forms of power.

Environmental impact of tidal power

The environmental impact of tidal power stations has not been fully explored for the simple reason that only one major power station exists. Although the potential environmental impacts would be specific to the individual site, a few generalizations can be made. A tidal power station would change the water level in an estuary, affecting patterns of vegetation growth. It would have an impact on the ecosystems of the shoreline and of the water. It would likely have an impact on the quality of the water in an estuary; for example, it could change the cloudiness of the water, which in turn could affect the types of fish that could live in the water. This which would in turn have an effect on birds that feed off the fish. Fish life would also be affected by a barrage unless a way was found to allow the fish to pass through. Further, a tidal station could change patterns of bird migration and reproduction.

Economic impact of tidal power

Because of the limited availability of suitable sites, only about 2 percent of potential tidal power can currently be harvested. The potential amounts to 3,000 gigawatts (*giga-*, meaning billion) of electricity, so roughly 60 gigawatts could actually be produced with current technology. The economic impact to tidal electricity would likely be local. For instance, it is estimated that a tidal power station on England's Severn River could produce up to ten percent of England's electricity.

Issues, challenges, and obstacles of tidal power

The chief issues facing tidal power are economic. The cost of building such a plant is high. However, once the plant is built, the energy it generates is essentially free, although the costs of maintaining the plant and distributing the power have to be included in cost estimates. The cost of such a plant would therefore be spread out over a period of thirty years or more, but finding initial funding is difficult. Also, because of limited experience with tidal power stations, their environmental impacts are not well understood. A final challenge is developing equipment that can withstand the harsh marine environment.

OCEAN WAVE POWER

Wave power is actually another form of solar power. As the sun's rays strike the Earth's atmosphere, they warm it. Differences in the temperature of air masses cause the air to move, resulting in winds. As the wind passes over the surface of the oceans, a portion of the wind's kinetic energy is transferred to the water, producing waves. These waves can travel essentially unchanged for enormous distances. But as they approach a shoreline and the water becomes shallower, their speed slows and they become higher. Finally, the wave collapses near shore, releasing an enormous amount of energy. It has been estimated that the amount of kinetic energy contained in a wave is up to 110 kilowatts per meter.

Capturing wave energy means that the kinetic energy of waves is converted into electrical power. In many respects, the technology is the same as it is with tidal and hydroelectric power. The kinetic energy turns a turbine attached to a generator, which produces electricity.

Current uses of ocean wave power

Scientists and engineers have devised hundreds of ways to capture wave power. The first, developed by a company called Wavegen, is being used at the world's only major wave power station in operation, the 500-kilowatt Land-Installed Marine-Powered

Energy Transformer (Limpet) on the island of Islay off Scotland's western coast. The basic design is called an oscillating water column (OWC). The water from a wave flows into a funnel and down into a cylindrical shaft. The rise and fall of the water in the shaft drives air into and out of the top of the shaft, where it blows past turbines, causing them to turn. In a sense, then, an OWC is a combination of hydropower and a windmill, with the "wind" consisting of air pressurized by the power of the wave. As with most other forms of hydropower, the turbines are attached to a generator, which produces electricity. In the case of Limpet, two turbines are in place. A chief advantage of this design is that the generators are not submerged in the water, making maintenance easier. Wavegen has built and tested a number of prototypes and in the early 2000s was constructing an OWC station on Pico Island in the Azores. It was anticipated that the plant would provide ten percent of the island's power requirement for its 15,000 people.

A second design is generally referred to as a wave-surge or focusing device. With these systems, sometimes called tapered channel or "tapchan" systems, a structure mounted on shore, which looks a little like a skateboard ramp, channels the waves and drives them into an elevated reservoir. As water flows out of the reservoir, it generates electricity in much the same way a hydroelectric dam does. A variation of this design was developed by a Norwegian company called WaveEnergy. This design consists of a series of reservoirs layered into a slope. WaveEnergy has also proposed attaching its design to old deep-sea oil-drilling platforms.

Engineers continue to work on other designs. One example that can be cited is the hosepump, which makes use of a type of hose called an elastomeric hose, the volume of which decreases as the hose is stretched in length. The hose is attached to a float that rides the waves on the ocean's surface, pulling it and relaxing it. This movement pressurizes seawater in the hose, which is then fed through a valve past a turbine attached to a generator. This is one example of the many ingenious devices with which scientists are experimenting. Many of these devices have fanciful names: the Mighty Whale, the Wave Dragon, Archimedes Wave Swing, WavePlane, Pendulor, and the Nodding Duck.

Benefits and drawbacks of ocean wave power

Like other forms of hydropower, wave power does not require the burning of fossil fuels, which can pollute the air, contributing to acid rain and global warming. The energy is entirely clean and endlessly

renewable. Further, in contrast to tidal power and thermal energy stations, which can be built in only a limited number of locations, wave power stations could be built along virtually any seacoast. Some of these devices could provide artificial habitats for marine life. They could also serve a secondary function as breakwaters.

The chief drawback of any onshore wave power station is the disruption caused to the natural environment by the presence of the station itself. OWC stations could potentially be noisy, although engineers continue to work on ways to dampen the noise they produce. A further drawback is that many of the technologies are new and untried, making it difficult to find funding to build the plants. In addition, these types of devices could cause navigational hazards for the shipping and fishing industries. Because of their location by the open ocean, these power stations could sustain severe damage from storms affecting the coastline, such as hurricanes.

Limpet 500, the world's first commercial-scale wave power station, generates 500 kilowatts of electricity, enough to power 300 homes. It lies on the coast of Islay, a Scottish Hebridean Island. As the wave moves into this partly-submerged hollow concrete chamber, air is forced out through a turbine-containing blowhole in its rear. When the wave falls, air is sucked back through the blowhole. Electricity is generated using a Wells turbine that rotates the same way despite the two-way air flow. *Martin Bond/Photo Researchers, Inc.*

Impact of ocean wave power

Wave power stations could impact the environment in a number of ways. Offshore or near-shore devices could change the flow of sediment, affecting marine life in unpredictable ways. Onshore devices could have an impact on, for example, turtle populations or other shoreline creatures that use the shorelines for nesting and breeding.

The economic impact of wave power is hard to calculate, but the potential impact is enormous. It is estimated that the total amount of wave energy that strikes the world's coastlines is about 2 to 3 million megawatts. In many locations throughout the world, the waves along one mile of coast contain the equivalent of 65 megawatts of power, or about 35,000 horsepower. Some experts say that if existing technologies were widely adopted, wave power could provide about 16 percent of the world's electricity needs. A large wave power station (100 megawatts) could provide power for as little as three to four cents per kilowatt-hour; a smaller station (1 megawatt) could provide power for seven to ten cents per kilowatt-hour. Both of these ranges include the cost of the plant's construction divided out over a period of years.

Issues, challenges, and obstacles of ocean wave power

As with other forms of water power, the chief obstacle is funding. Many wave-power technologies are unproven, particularly on a large scale, so it is difficult for developers to attract funding from private and governmental organizations. Another challenge is building equipment that is sturdy enough to withstand the harsh marine environment over long periods of time.

■ ■ ■

For More Information

Books

Avery, William H., and Chih Wu. *Renewable Energy from the Ocean.* New York: Oxford University Press, 1994.

Berinstein, Paula. *Alternative Energy: Facts, Statistics, and Issues.* Phoenix, AZ: Oryx Press, 2001.

Boyle, Godfrey. *Renewable Energy,* 2nd ed. New York: Oxford University Press, 2004.

Cuff, David J., and William J. Young. *The United States Energy Atlas,* 2nd ed. New York: Macmillan, 1986.

Howes, Ruth, and Anthony Fainberg. *The Energy Sourcebook: A Guide to Technology, Resources and Policy.* College Park, MD: American Institute of Physics, 1991.

Periodicals

Freeman, Kris. "Tidal Turbines: Wave of the Future?" *Environmental Health Sciences* (January 1, 2004): 26.

Valenti, Michael. "Storing Hydroelectricity to Meet Peak-Hour Demand." *Mechanical Engineering* (April 1, 1992): 46.

Web sites

O'Mara, Katrina, and Mark Rayner. "Tidal Power Systems." http://reslab.com.au/resfiles/tidal/text.html (accessed on September 13, 2005).

"Tidal Power." University of Strathclyde. http://www.esru.strath.ac.uk/EandE/Web_sites/01-02/RE_info/Tidal%20Power.htm (accessed on September 13, 2005).

Vega, L. A. "Ocean Thermal Energy Conversion (OTEC)." http://www.hawaii.gov/dbedt/ert/otec/index.html (accessed on September 13, 2005).

Weiss, Peter. "Oceans of Electricity." *Science News Online* (April 14, 2001). http://www.science news.org/articles/20010414/bob12.asp (accessed on September 13, 2005).

Wind Energy

INTRODUCTION: WHAT IS WIND ENERGY?

The word "windmill" for many people brings to mind the Netherlands, whose countryside for centuries has been dotted with thousands of windmills. Windmills represent an early technical skill or ingenuity (inventiveness) that seemed to be lost during the industrial revolution, when fossil fuels replaced wind and running water as the most widely used energy sources. Some people of the twenty-first century support a return to greater reliance on the wind that powers windmills, chiefly because wind power is clean and endlessly renewable.

Historical overview

The first written record of a windmill is in a Hindu book from about 400 BCE (before the common era). About four hundred years later, the Greek inventor Hero of Alexandria devised a wind-driven motor he used to provide air pressure to operate an organ. From about 400 CE (common era), there are references to prayer wheels driven by wind and water in the Buddhist countries of central Asia. These devices were handheld windmills that contained prayers and religious texts on rolls of thin paper wound around an axle. Individuals could access the prayers whenever they wanted (the thought was increasing the speed of the spinning prayer wheels strengthened the prayers). Early devices used the power of the wind, but it was not until much later that wind power was developed as a way to do work.

Some historians believe that the earliest true windmills—that is, windmills built to do work—were built in China two thousand years ago, but no records exist. The first recorded references to true windmills date from seventh-century Persia, later called Iran, particularly the province of Sijistan, which became Afghanistan.

Words to Know

Anemometer A device used to measure wind speed.

Coriolis force The movement of air currents to the right or left caused by Earth's rotation.

Drag The slowing force of the wind as it strikes an object.

Kilowatt-hour One kilowatt of electricity consumed over a one-hour period.

Kinetic energy The energy contained in a mass in motion.

Lift The aerodynamic force that operates perpendicular to the wind, owing to differ-ences in air pressure on either side of a turbine blade.

Nacelle The part of a wind turbine that houses the gearbox, generator, and other components.

Rotor The hub to which the blades of a wind turbine are connected; sometimes used to refer to the rotor itself and the blades as a single unit.

Stall The loss of lift that occurs when a wing presents too steep an angle to the wind and low pressure along the upper surface of the wing decreases.

Wind farm A group of wind turbines that provides electricity for commercial uses.

During the reign of the Muslim caliph 'Umar I (633–44), windmills were constructed primarily to obtain water for irrigating crops and grinding grain. These working windmills may have been imported into China from the Middle East by Genghis Khan (1162–1227), the Mongol conqueror of much of what is now Iran and Iraq (1216–23). The first reference to a Chinese windmill dates from the year 1219, when a statesman named Yehlu Chhu-Tshai docu-mented construction of one. Windmills became widely used along the coasts of China during this period.

The design of these seventh-century windmills, some of which survive in Iran and Afghanistan, was the reverse of modern wind-mills. In modern windmills the axle is horizontal and is positioned at the top of the windmill. In early Middle Eastern windmills the blades that turned in the wind were enclosed in a chamber at the bottom of the windmill. The blades were attached to a vertical axle, which was attached to a millstone above. The early windmills, which are still used, could grind a ton of grain per day and generate about one-half the power of a small car.

Windmills in Europe

During the Crusades, which took place over a two-hundred-year period beginning in 1095, European conquerors of Palestine probably became familiar with Middle Eastern windmills and imported the technology back to Europe. The first documented reference to a

This windmill, seen in the Netherlands, is typical of what many people envision for windmills. © *Royalty-Free/ Corbis.*

European windmill dates to 1105 in France, the home of most of the early crusaders. A similar reference is made to a windmill in England in 1180. Both of these windmills were built to pump water to drain land.

For reasons that are unknown, the Europeans mounted the windmill blade on a horizontal axle rather than a vertical one. They may have adopted the design from water wheels, which by this time were being mounted on horizontal axles (poles around which an object rotates). Some of the windmills from this period were able to lift more than 16,000 gallons (60,566 liters) of water per hour, using augers (a type of screw) that raised the water from lower levels to higher levels, where the water could be sent into channels. The augers acted like spiral staircases that carried the water up as the windmills turned.

Al-Dimashqi Describes a Windmill

In the thirteenth century, the Arab historian al-Dimashqi (1256–1327), described a windmill:

> When building mills that rotate by the wind, they proceed as follows. They erect a high building, like a minaret, or they take the top of a high mountain or hill or a tower of a castle. They build one building on top of another. The upper structure contains the mill that turns and grinds, the lower one contains a wheel rotated by the enclosed wind. When the lower wheel turns, the mill stone above also turns. . . . Such mills are suitable on high castles and in regions which have no water, but have a lively movement of the air.

These windmills were often arranged in what were called gangs, meaning that they were arranged in rows so that water could be drained in stages, especially from lower to higher levels.

Because much of the Netherlands is below sea level, the Dutch made extensive use of windmills to drain land and to grind grain. By the fourteenth century the Dutch had introduced or adopted a number of technologies, such as post mills and tower mills. The post mill consisted of a four-bladed mill mounted on a central vertical post or shaft. Wooden gears transferred the power of the shaft to a grindstone. The grindstone turned to make grain into flour. The tower mill, which originated along the Mediterranean seacoast in the thirteenth century, consisted of a post mill mounted on top of a multistory tower. This tower housed the grinding machinery and had rooms for grain storage and other milling functions as well as living quarters in the bottom story. The tower mill is the type most often seen in pictures of Dutch windmills.

A major concern of windmill operators was to make sure that the mill was positioned correctly in relation to the wind. This task was done with a large lever at the back of the windmill that was pushed to move the windmill blades toward the wind. The blades were made of lattice frames over which canvas sails were stretched. By 1600, windmills were in such widespread use in Holland that the bishop of Holland, seeing a chance to increase funds for the church, declared an annual tax on windmill owners.

Also by that time the basic technology of windmills was in place. It remained for engineers and inventors to find ways to increase efficiency, primarily by coming up with new designs for windmill blades. Some of these designs included improvements in the blade's

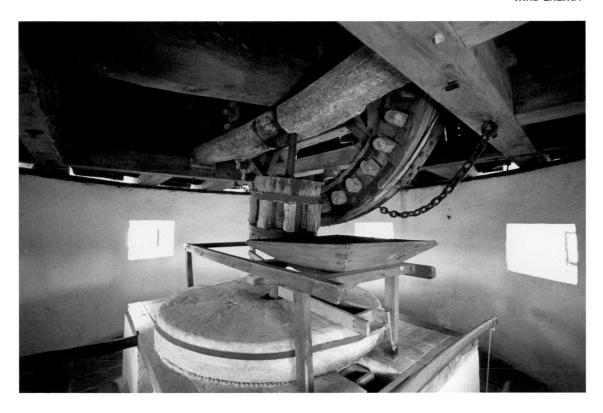

camber, or the outward curve of the blade from its leading edge (the edge first struck by the wind) to its trailing edge. Other experiments were conducted to find the best location for the blades spar, or the long piece of a blade; its center of gravity; and the correct amount of twist in the blade. One of the most prominent millwrights (mill builders) during the period, Jan Adriaanzoon Leeghwater (1575– 1650), experimented with these matters. Largely through his efforts, about twenty-six lakes in the Netherlands were drained.

By the end of the nineteenth century, at least 30,000 windmills were operating in Europe. These windmills were used not only to pump water and grind grain but also to power sawmills and for other industrial uses, including processing agricultural products such as spices, cocoa, dyes, paints, and tobacco.

Windmills in North America

In the seventeenth and eighteenth centuries, the Dutch migrated to the American colonies in large numbers. They brought with them the technology for constructing windmills, and many Dutch-style windmills were built throughout New York and New England, where they worked well in the relatively gentle eastern winds.

Interior of a windmill, in Spain, showing the wooden gears that were powered by the wind. © *Corbis.*

An illustration depicting an early wind mill (around 1430), with an automatic elevator for lifting flour bags. The post was designed to turn in the direction of the wind. *Bettmann/Corbis.*

In the nineteenth century, American settlers moved westward and onto the Great Plains. The settlers wanted to harness the power of the wind to irrigate the land and water their cattle. However, on the plains a fundamental design flaw in Dutch windmills became apparent: The slow-moving blades were too fragile for the strong winds that swept across the prairies in places such as Kansas and Nebraska. As soon as they were hit with high winds, these windmills fell apart.

What's in a Name?

One project that Jan Adriaanzoon Leeghwater started in Holland was a drainage plan to protect Amsterdam and Leiden from the Haarlem Meer, a lake that was growing each year and threatening to flood the cities. The project that he began in 1643 was so large that it was not completed until 1852. One of the three pumping stations still operating in the early twenty-first century was named after Leeghwater. The engineer's life course may have been set the day he was born. In Dutch "Leeghwater" means "empty water."

Back in New England, a designer named Daniel Halladay (1826–?) patented a design that could withstand the high winds of the plains. His company, the Halladay Windmill Company, began building windmills with the new design in 1854. The chief improvement Halladay made was to use numerous blades, rather than the four blades that were common on New England windmills. The new windmills also had a tail that would orient them to the wind, and they had hinged blades that would fold up in high winds so that they would not fall apart. In 1857 Halladay's company began doing business as the U.S. Wind Engine and Pump Company.

In about 1870 windmill manufacturers made another improvement when they began using steel rather than wood in the manufacture of blades. These blades were stronger but also could be curved, making them much more efficient than the flat wooden blades in use up to this time. In 1886 the inventor Thomas Perry designed a more aerodynamic blade, a blade that gets the most power from the wind and a design that continues to be used in the early twenty-first century.

Halladay's company, along with numerous competitors, sold thousands of windmills. Many windmills were sold to farmers and ranchers, but another industry emerged as a major customer. The railroads needed large amounts of water for their steam engines at their many stops across the plains and on to the West Coast. Windmill-powered pumps pumped water into tanks at the side of the railroad tracks. Trains could stop at each tank and get water enough to continue the journey to the next tank.

Another major improvement occurred in 1915, when the Aeromotor Company designed an enclosed, self-lubricating gearbox. Until then, the open gears of windmills had to be lubricated every week, often by horse-mounted cowboys who rode out with their saddlebags

Windmills are still used today by farmers and ranchers to pump water for family and livestock use. © 2005 Kelly A. Quin.

packed with bottles filled with oil. In windmills with the Aeromotor gearbox, the gears had to be oiled only about once a year.

About one million windmills made by about 300 companies were built in the United States between 1850 and 1970. Although most of these windmills were small, and used on family farms primarily to pump water, others were large, with blades up to 26 feet (8 meters) long. These were purchased mainly by the railroads for their system of track-side water tanks.

Electrification

The next step in the development of wind energy was electrification. Until the late nineteenth century, all windmills produced only mechanical power for pumping or grinding. With the emergence of electricity, designers and engineers quickly recognized that windmills could be attached to electric generators and that the power they produced could be used for heating and lighting.

The first windmill used to generate electricity on a large scale was built in 1888 by Charles F. Brush (1849–1929) in Cleveland, Ohio. Its rotor, which consisted of 144 blades, was almost 56 feet (17 meters) in diameter. The rotor includes the hub and the blades that are attached to it. Brush's major technical challenge was to find a way for the windmill's rotor to produce the 500 revolutions per minute he needed for the generator to operate. Brush designed a step-up gearbox (a series of parts that transmitted motion from one part of the machinery to another) in a fifty-to-one ratio. This meant that for every turn of the rotor, the operational parts of the generator turned 50 times. During the 20 years it was in operation, the Brush machine produced about 12 kilowatts of power, which Brush stored in batteries in his nearby mansion.

From 1890 to 1930 the windmill industry in the United States boomed. Spurring the boom was the prominent place given to electric windmills at the World's Columbian Exposition in Chicago in 1893, where they were used to generate power to light the fairgrounds after dark. Electric lights were not common in 1893 homes; most still used gaslights. So people were amazed that a cheap source of power could make this new marvel available to them, even if they lived out in the country. However, the windmill industry soon collapsed after the U.S. Rural Electrification Administration, or REA, was established. This government program was one of many created to help the nation overcome the effects of the Great Depression (1929–1941). The REA provided partial federal funding for electricity to homes and farms in rural areas, much of it produced by hydroelectric dams. If these hard-to-reach places could now get inexpensive electrical service from the government, then they no longer needed windmill-generated power.

Decline and revival

From the 1930s to the 1970s in the United States coal and oil remained relatively inexpensive, and little interest was shown in harnessing the wind to meet the need for electricity. In Russia, however, a 100-kilowatt wind generator was built in Balaclava in 1931. Mounted on a tower 100 feet (33 meters) high, the rotor was 100 feet in diameter and produced power when the wind speed exceeded 25 miles (40 kilometers) per hour. The wind generator supplied this energy to a steam power station 20 miles (32 kilometers) away. The turbine did not

Watts, Kilowatts, and Kilowatt-hours

Electric output is generally measured in watts, named after the Scottish inventor James Watt (1736–1819). A watt is 1/746th of one horsepower (the power of one horse pulling). Because 1 watt is a small amount, power is generally measured in kilowatts, or thousands of watts. Large power-generating stations often measure power output in megawatts, or millions of watts.

By itself a wattage figure does not indicate how much power is being consumed. A 100-watt lightbulb needs 100 watts to operate, but more power is consumed if the light is left on for an hour than if it is left on for a minute. The term "kilowatt-hour" takes into account the time dimension. If a 100-watt bulb is left burning for 10 hours, 1 kilowatt-hour of electricity has been consumed. A typical family in the United States uses about 10,000 kilowatt-hours of electricity each year.

last very long because the blades were made of old roofing metal and the gears were made of wood. During one year of operation, however, the wind generator produced 279,000 kilowatt-hours of power.

From the mid-1930s until 1970 commercial-sized wind generators were built in Denmark, England, Germany, and France. These countries were left with shortages of fossil fuels and most everything else because of the destruction left by World War II (1939–1945). The development of wind power in Europe filled some of the need for electricity that was not being filled by fossil fuels. In Denmark, for example, a 200-kilowatt wind generator was built and operated until the early 1960s. Denmark led the way in wind-power generation in terms of the percentage of electricity that was wind generated, about 20 percent.

Although Europe was leading the way, the largest commercial-grade wind generator was located on Grandpa's Knob, a 2,000-foot-high (610 meters) hill near Rutland, Vermont. It was called the Smith-Putnam wind turbine after its designer, Palmer C. Putnam, and the company that provided the money to build it, the S. Morgan Smith Company of New York. The generator was built over a two-year period beginning in 1939. The 175-foot-diameter (53 meters) rotor produced an enormous 1.25 megawatts of power during the four years it was in operation. The Smith-Putnam turbine stopped operating when metal fatigue caused

some of the blades and bearings to break. Replacements could not be found because metals and other materials were being used by the military to build weapons to fight World War II. Although the Smith-Putnam turbine was not a long-term economic success, it was considered a technical success because it produced a lot of electrical power while it was working.

During the years following World War II, several wind energy designs were built and tested. In England the Enfield-Andreau wind turbine, built in St. Alban's in the 1950s, had a 79-foot (24 meters) rotor that produced 100 kilowatts of power. A unique feature of this turbine was that its hollow propeller blades acted as air pumps for transmitting power from the rotor to the generator.

In Denmark the Gedser wind turbine was built in 1957, and its 79-foot blades produced about 400,000 kilowatt-hours per year until the turbine was shut down in 1968. Also during the 1950s, two large machines were built in France. One produced 130 kilowatts and the other 300 kilowatts. In Germany the Hütters wind turbine achieved great efficiency by producing 100 kilowatts of power in only 18-mph (29 kph) winds. Earlier systems needed higher wind speeds.

Wind turbines capture the kinetic energy of wind with blades shaped much like airplane propellers. These blades are attached to a tower that rises at least 100 feet (30 meters) above the ground. © *George D. Lepp/ Corbis.*

The Coriolis force

The Coriolis (kawr-ee-OH-luhs) force, some-times called the Coriolis effect, is named after the French mathematician Gaspard-Gus-tave de Coriolis (1792–1843). The principle behind the Coriolis force is that because Earth rotates, any movement in the Northern Hemisphere is diverted to the right, if observed from a fixed position on the ground. In the Southern Hemisphere, the movement is to the left. This means that wind tends to rotate counterclockwise around low-pressure areas in the Northern Hemisphere and clock-wise in the Southern Hemisphere.

The Coriolis force has a major effect on pre-vailing wind patterns throughout the world. As equatorial air heats, rises, and moves toward the poles, expansion of the air cre-ates low pressure. Cooler air from the poles flows in behind the warmer air to equalize the pressure. At about 30 degrees latitude north and south, the Coriolis force prevents air from moving much farther toward the poles,

because the warmer air encounters a high-pressure area of cooler, sinking air. Because of the diversion of the air caused by Earth's rotation, prevailing winds generally blow in the following directions:

Latitude	Direction
90°–60°	N Northeast
60°–30°	N Southwest
30°–0°	N Northeast
0°–30°	S Southeast
30°–60°	S Northwest
60°–90°	S Southeast

© 1997-2003 Danish Wind Industry Association. Reproduced by permission. *Thomson Gale*

The Coriolis force does not explain wind direction in all places at all times. Local factors also determine the speed and direc-tion of the wind. A good example is a sea breeze. Land masses warm faster in the sun than water does. This means that the air over land expands and rises faster than the air over the sea. As the land air rises,

During the 1970s it seemed as though the United States was ready to make the necessary investments to develop wind power. In 1973 the country was affected by the Arab oil embargo. Countries that normally sold oil to the United States were refusing to do so. This served as a warning to the nation that it was too dependent on foreign oil, which could be cut off at any moment. In 1974 the U.S. Federal Wind Energy Program was established. Over the next decade scien-tists from U.S. agencies such as the National Aeronautics and Space Administration (NASA) and the U.S. Department of Agriculture built and tested at least thirteen different wind turbine designs, ranging in output from 1 kilowatt to 3.2 megawatts. Major efforts were made to develop more efficient rotor designs. Many of these designs were successful, and engineers learned to design better ones.

However, by the late 1980s it was becoming more and more difficult to attract funding for wind energy efforts. Many people remained unconvinced that wind power could ever provide more than small

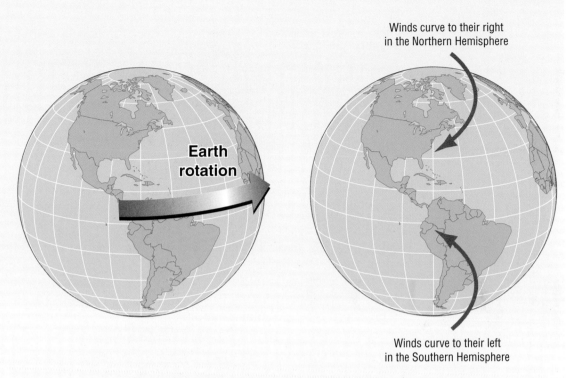

Winds curve to their right
in the Northern Hemisphere

Earth
rotation

Winds curve to their left
in the Southern Hemisphere

Illustration of the Coriolis effect in the Northern and Southern Hemispheres as the Earth globe rotates. *Thomson Gale.*

the sea air flows in behind it, causing wind to blow onshore. At night, the process is reversed, and wind tends to blow offshore, that is, from land out to sea. Mountain ranges also play tricks with the wind, diverting it in different directions.

amounts of electricity for local use. Since that time research on wind technology has been conducted in the United States largely by the National Wind Technology Center near Boulder, Colorado.

HOW WIND ENERGY WORKS

In everyday discussions of alternative forms of energy, most people make a distinction between wind power and solar power. From one point of view, however, this distinction is unnecessary because the wind that powers wind turbines is itself a form of solar power.

Earth absorbs overwhelming amounts of energy from the sun: 1.74×10^{17} kilowatt-hours, or 174,423,000,000,000 kilowatts every single hour of the day. Although the oceans and land masses absorb a great deal of this energy, much is absorbed by the atmosphere (the whole mass of air surrounding Earth).

The energy from the sun does not strike Earth evenly. Air around the equator absorbs more energy than the air above the poles. This difference causes air, a fluid much like water, to move in currents. Air, like any substance, expands when it is warmed and contracts when it is cooled. Warm air, because it is less dense than cool air, is lighter, so it rises, much like a less-dense piece of wood rises to the top of more-dense water. This effect can be seen by looking at the hot air above a fire, which seems to shimmer as it expands and moves upward, carrying smoke and ash with it. Cold air, because it shrinks, is denser than surrounding warm air, so it sinks. This property explains in part why a freezer generally operates more efficiently when it is placed at the bottom of a refrigerator rather than at the top and why the basement is generally colder than the upper levels of a house.

As warm air rises, colder, heavier air flows in to replace it, causing a current of air—in other words, wind. Earth's rotation also plays a role in wind production. If Earth did not rotate, air heated at the equator would rise only about 6 miles (10 kilometers) into the atmosphere and flow toward the North Pole and the South Pole, where it would cool, sink, and return to the equator. Earth's rotation allows winds to circulate in more or less predictable patterns across the Northern Hemisphere and Southern Hemisphere. These winds contain huge amounts of kinetic (kuh-NET-ik) energy, or the energy contained in any fluid body in motion. About two percent of the solar energy that strikes Earth is converted into wind. For various reasons, including the revolution of Earth and features of its terrain, some parts of Earth have more wind than others.

The southeastern United States has relatively little wind on a steady basis, so this region is generally not considered a good place to place wind turbines. In addition, the storminess in the Southeast would leave wind turbines vulnerable to damage from high winds, during hurricane season, for example. The Rocky Mountain states experience a great deal of wind on a consistent basis, making them better candidates for wind power. The best places to build the turbines are North Dakota, Texas, and Kansas, which by themselves could provide all of the electricity needed in the United States, according to a 1991 U.S. Department of Energy wind resource report.

According to the Battelle Pacific Northwest Laboratory, the top twenty states and the amount of wind power they could produce, measured in billions of kilowatt-hours per year, are as follows:

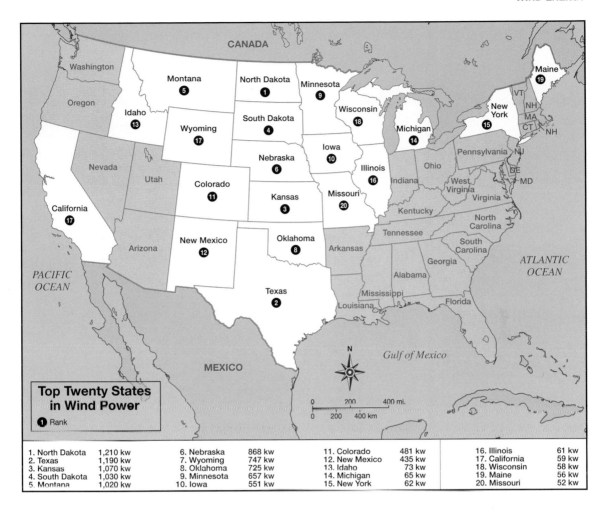

Top Twenty States in Wind Power

1 Rank

1. North Dakota	1,210 kw	6. Nebraska	868 kw	11. Colorado	481 kw	16. Illinois	61 kw
2. Texas	1,190 kw	7. Wyoming	747 kw	12. New Mexico	435 kw	17. California	59 kw
3. Kansas	1,070 kw	8. Oklahoma	725 kw	13. Idaho	73 kw	18. Wisconsin	58 kw
4. South Dakota	1,030 kw	9. Minnesota	657 kw	14. Michigan	65 kw	19. Maine	56 kw
5. Montana	1,020 kw	10. Iowa	551 kw	15. New York	62 kw	20. Missouri	52 kw

According to the American Wind Energy Association, by the end of 2004 wind facilities in thirty U.S. states were generating a total of 6,740 megawatts of electricity, enough to provide power for about 1.6 million homes.

The states leading the way were these:

California: 2,096 megawatts

Texas: 1,293 megawatts

Iowa: 632 megawatts

Minnesota: 615 megawatts

Wyoming: 285 megawatts

The largest wind farms, or large facilities with numerous turbines, operating in the United States were the following:

Stateline, Oregon-Washington: 300 megawatts

An Assessment of the Available Windy Land Area and Wind Energy Potential in the Contiguous United States, Top Twenty States in Wind Power, August, 1991, PNL# 7789. Reproduced by permission. *Thomson Gale.*

Commercial wind power usually is generated at wind farms rather than from single turbines. The largest wind farm in the United States is the Stateline Wind Energy Center, located on the Vansycle Ridge, which runs along the Columbia River on the Washington-Oregon border. © *Russell Munson/Corbis.*

© 2004 American Wind Energy Association. Reproduced by permission. *Thomson Gale.*

King Mountain, Texas: 278 megawatts

New Mexico Wind Energy Center, New Mexico: 204 megawatts

Storm Lake, Iowa: 193 megawatts

Colorado Green, Colorado: 162 megawatts

High Winds, California: 162 megawatts

The countries that led the world in wind power production in 2004 were as follows:

World Leaders in Wind Capacity, December 2004

Country	Capacity in Megawatts
Germany	116,629
Spain	8,263
United States	6,740
Denmark	3,117
India	3,000
Italy	1,125
Netherlands	1,078
United Kingdom	888
Japan	874
China	764

Table 3 World Leaders in Wind Capacity, December 2004

Source: American Wind Energy Association, http://www.awea.org/faq/tutorial/wwt_statistics.html

CURRENT AND FUTURE TECHNOLOGY

Throughout the twentieth century, engineers experimented with various rotor designs. One was called the Darrieus windmill, named after the person who invented it in the 1920s. Rather than using blades that look like airplane propellers, the Darrieus windmill looks more like a giant eggbeater, with thin blades connected at the top and bottom of a vertical shaft. The Darrieus windmill has the advantage of working no matter which way the wind is blowing. In addition, generators can be mounted at the bottom rather than the top.

The most common type of windmill in the early twenty-first century was called the vertical-axis wind turbine, which had airplane propeller-type blades mounted at the top of a tall tower. This windmill, called the MOD-2, was designed by NASA. Each MOD-2 was mounted on a 200-foot-tall (61 meters) tower. The blades were up to 150 feet (46 meters) long. The MOD-2 could produce about 2,500 kilowatts of power in a 28-mph (45 kph) wind. Other wind turbine rotors may be larger, but their fundamental design owes much to the design of the MOD-2.

The technology of wind-power generation is well-developed. Although refinements in blade configuration and other factors probably can be made, the technology is cost-effective and sound. The major challenge for the future is harnessing the technology on a big enough scale to provide power to large numbers of users.

BENEFITS AND DRAWBACKS OF WIND ENERGY

The chief benefits of wind power are that it is clean, safe, and endlessly renewable. The fuel that powers wind turbines is free, so its price to utility companies does not vary. Wind power does have a number of drawbacks. Wind speed does not remain constant, so the supply of power may not always be the same as demand from consumers. Because many of the best locations for wind turbines are far from urban areas, there are problems with distributing the energy.

Environmental impact of wind energy

Wind power is clean and renewable, but it also raises environmental concerns. Wind power farms require large stretches of land or have to be placed in environmentally sensitive areas such as deserts or on ridgelines. Many people consider wind farms

The Darrieus wind turbines have the advantage of working no matter what direction the wind is blowing. *U.S. Department of Energy, Washington D.C.*

unsightly, a form of visual pollution. A major environmental concern is the effect of wind farms on patterns of bird migration. Many birds have been killed by flying into wind turbine blades.

Economic impact of wind energy

The cost of generating electricity with wind power has steadily decreased. Wind-power electricity can be generated for about four to six cents per kilowatt-hour, making wind power competitive with other forms of generation of electricity.

Societal impact of wind energy

The societal impact of wind power is similar to that of many other renewable fuels. About two billion people worldwide do not have electricity. Many of these people live in areas where connecting them to the power grid would be extremely expensive. Wind power may be an alternative way to provide power to these people, improving their quality of life.

Wind power also may reshape the way people think about electricity and their place in a nation's power distribution system. Most electric power is provided by huge facilities, which often are far from the consumer's home or business. Wind power, at least for the near future, is likely to be generated closer to home, in communities and even at the neighborhood level. As fossil fuels become increasingly more expensive and eventually are depleted, alternative energy, including wind, solar, tidal, and wave power generated locally, may contribute to a sense of people belonging to communities rather than to large, anonymous societies. Decisions about power supplies and distribution would be made close to home in response to local needs.

Wind power can have harmful effects on the environment. Some environmentalists are concerned about soil erosion, bird safety, and noise pollution.
© PICIMPACT/Corbis.

WIND TURBINES

In the early twenty-first century, wind turbines are mainly used to produce electricity. Some turbines are on wind farms and contribute electricity to the power grid for commercial use. Remote areas also use turbines, providing electricity to small villages that are too far away from the transmission lines of the commercial areas. Turbines have other uses besides producing electricity, such as pumping water, and ice making. Near oceans there is some use of wind turbines to help remove the salt from the ocean water.

How wind turbines work

The technology of wind turbines is simple. Wind turbines capture the kinetic energy of wind with, in most cases, two or three blades shaped much like airplane propellers. These blades are attached to a tower that rises at least 100 feet (30 meters) above the ground. At this height air currents tend to be stronger but less turbulent than they are at ground level. When the wind strikes the blade, the angle and configuration of the blade form a pocket of low pressure on the downwind side of the blade. This low pressure sucks the blade into movement, causing the rotor to turn. Force is added by the high pressure on the upward side of the blade. In aerodynamic theory, this property is called lift. If the blade is designed correctly, lift is stronger than drag, or the slowing force exerted by the wind on the front of the blade.

In wind turbines lift and drag work together to make the entire mechanism spin like a propeller. In earlier windmills drag rather than lift was the force that turned the blades. The process is the opposite of that of a fan. With a fan electricity is used to make wind. With a wind turbine wind is used to produce electricity. The turning rotor of a wind turbine is connected to a shaft, which is connected to an electric generator. Power can be distributed to users over the electric grid in exactly the same way any other electric power is distributed.

The most important feature in the operation of wind turbines is lift. To achieve lift, wind turbine designers have borrowed technology from aircraft designers. In cross-section an airplane wing looks like an irregularly shaped teardrop. The shape is irregular because the wing's bottom is slightly flatter than the top, which is more curved. When a plane flies, its wings slice through the air, creating wind. Because of the curve of the upper surface of the wing, the air has to flow faster to get around the wing. At the same time, the air flows at a lower speed along the bottom surface of the wing. Because of the difference in speed, the air above the wing is less dense; that is, the air pressure is lower than the pressure of the air below the wing. This difference in pressure creates lift perpendicular to the direction of the moving air, allowing the plane to fly. The same principle applies to turbine blades.

Unlike airplane wings, wind turbine wings are almost always twisted. The reason they are twisted has to do with another aerodynamic principle, stall. When an airplane wing is tilted back, the wind continues to flow smoothly along the bottom surface, but along the top surface, because of the steeper angle presented to the wind, the air no longer sticks to the wing but swirls around in a circle above it. The result of this swirling is the loss of the low pressure along the upper surface of the wing. Without this low pressure, the plane has no lift and drops like a rock.

Unlike airplane wings, wind turbine blades are constantly rotating, and the speed of the rotation differs along the entire length of the blade. At the precise geometric center, the speed of rotation is zero. This speed steadily increases along the length of the blade until at the tip the blade can be moving hundreds of feet (meters) per second. This rotation changes the direction at which the wind hits the blade all along its length. In effect, the angle at which the wind hits the blade would be different at each point along the blade if the blade were not twisted. When the blade is twisted, the angle at which the wind hits the blade is the same at each point, and stall is eliminated under normal wind conditions. Excessively high wind speeds can damage rotors, however, so engineers have designed blades that stall when the wind is too strong, and the rotor stops spinning.

Wind turbines come in two configurations. One, called a vertical-axis turbine, looks much like an oversized eggbeater. The axis of the turbine is positioned vertically, and the blades are connected to the axis at the top and the bottom. This configuration has one primary advantage: The turbine does not have to be faced into or away from the wind, so it operates no matter which way the wind is blowing, and it does not have to be repositioned to accommodate changes in wind direction.

The other configuration, the horizontal-axis turbine, is much more commonly used. With this style, the axis is parallel to the ground on a tower, and the blades, which look like airplane propellers, are perpendicular to the axis. This type of wind turbine looks like a pinwheel.

The Mathematics of Wind Energy

Three factors determine how much energy the wind can transfer to a wind turbine: the density of the air, the area of the rotor, and the speed of the wind. The first factor is air density. Any moving body contains kinetic energy. The amount of this energy is proportional to the body's mass or weight. A truck hurtling down the road at 50 miles per hour (80 kilometers per hour) has more kinetic energy, and consumes more gasoline, than a subcompact car traveling at the same speed. With wind the amount of kinetic energy depends on the density of the air. Heavy air contains more energy than light air. When the atmospheric pressure is normal and the air temperature is 59°F (15°C), air weighs 1.225 kilograms per cubic meter (0.076 pounds per cubic foot). Humid, or damp, air is denser than dry air, so it weighs more. Air at high altitudes, such as in mountain regions, is less dense, so it is lighter.

The second factor that determines the amount of energy the wind can transfer to a wind turbine is the area of the rotor. The diameter of a 1,000-kilowatt wind turbine is 54 meters (177 feet). Rotor diameters can vary with designs, but this diameter is typical. The area over which a rotor of this size operates is 2,300 square meters (24,757 square feet). As the diameter of a rotor increases, the increase in the area it covers increases with the square of the diameter. Thus, doubling the size of a turbine allows it to receive four times as much energy, or $2^2 = 2 \times 2$.

The third factor that determines how much energy the wind can transfer to a wind turbine is the speed of the wind. The relation between wind speed and energy is cubic. In other words, when the speed of the wind doubles, the amount of energy increases eight times, or $2^3 = 2 \times 2 \times 2$.

When the three factors are put together, the formula used to calculate the amount of wind energy available at a given site is $P = 0.5 \, \rho \, v^3 \, \pi \, r^2$ where P equals power measured in watts; ρ or the Greek letter rho (ROH), equals the density of dry air in kilograms per cubic meter (1.225); v equals the speed of the wind measured in meters per second; π, or the Greek letter pi (PYE), equals 3.14159; and r equals the radius, or half the diameter, of the rotor in meters.

A wind turbine has the following components:

- Rotor and blades. The rotor is the hub around which the blades are connected. Often, however, "rotor" is used to refer to the hub and the blades as a single unit. The rotor is the key component, because it translates the wind's kinetic energy into torque (TORK), or turning power.

- Nacelle (nuh-SELL), or the enclosure that houses the turbine's drive train, including the gearbox, the yaw mechanism, and the electric generator. The gearbox connects a low-speed shaft to a high-speed shaft. This mechanism can increase the speed of the shafts by a factor of as much as fifty to one, meaning that the high-speed shaft turns fifty times faster than the low-speed shaft. The yaw mechanism automatically senses the direction of the wind and rotates the rotor to keep it facing into the direction of the wind.

- Tower, or the support for the rotor and drive train.
- Electric equipment such as controls, cables, and an anemometer (an-uh-MAH-muh-tuhr)

Blades come in various sizes and have tended to grow over the years. In the early 1980s, a typical blade was likely to be 33 feet (10 meters) long, and such a wind turbine could generate about 45 megawatt-hours per year. By 1990 the typical blade measured 89 feet (27 meters) and could produce 550 kilowatt-hours per year. In the early twenty-first century blades as long as 233 feet (71 meters) can generate 5,600 megawatt-hours per year.

Building a wind turbine is far more than simply a matter of finding a field or mountaintop where the wind is blowing and plopping one down. Engineers give a great deal of attention to finding the proper site for a wind turbine. The main factor they consider is the average speed of the wind over an extended time. Using a device called a wind-cup anemometer, which looks like three or four ice-cream scoops arranged in pinwheel fashion, engineers take extensive measurements of wind speed over a long time.

Wind speed measurements have to be precise. If engineers overestimate the amount of wind, the power output of the turbine can be reduced considerably. If, for example, wind is believed to average 10 miles (16 kilometers) per hour but is only 9 miles (14 kilometers) per hour, the power output of the turbine is reduced 27 percent. If the wind speed is only 8 miles (13 kilometers) per hour, the power output is 41 percent less than expected. If the wind speed is higher than believed, power output increases. If the wind speed is 11 miles (18 kilometers) per hour, the power generated increases 33 percent. If the wind speed is much higher than expected, the equipment may be too small and too fragile for the site.

In addition to wind speed when looking for a place for a wind turbine, engineers consider factors such as wind hazards, characteristics of the land that affect wind speed, and the effects of one turbine on nearby turbines in wind farms. The following factors are important:

- Hill effect. When it approaches a hill, wind encounters high pressure because of the wind that has already built up against the hill. This compressed air rises and gains speed as it approaches the crest, or top, of the hill. Siting wind turbines on hilltops takes advantage of this increase in speed.

- Roughness, or the amount of friction that Earth's surface exerts on wind. Oceans have very little roughness. A city or a forest has a great deal of roughness, which slows the wind.

- Tunnel effect, or the increase in pressure air undergoes when it encounters a solid obstacle. The increased air pressure causes the wind to gain speed as it passes between, for example, rows of buildings in a city or between two mountains. Placing a wind turbine in a mountain pass can be a good way to take advantage of wind speeds that are higher than those of the surrounding air.

- Turbulence, or rapid changes in the speed and direction of the wind, often caused by the wind blowing over natural or artificial barriers. Turbulence causes not only fluctuations in the speed of the wind but also wear and tear on the turbine. Turbines are mounted on tall towers to avoid turbulence caused by ground obstacles.

- Variations in wind speed. During the day, winds usually blow faster than they do at night, because the sun heats the air, setting air currents in motion. In addition, wind speed can differ depending on the season of the year. This difference is a function of the sun, which heats different air masses around Earth at different rates, depending on the tilt of Earth toward or away from the sun.

- Wake. Energy cannot be created or destroyed. As wind passes over the blades of a turbine, the turbine seizes much of the energy and converts it into mechanical energy. The air coming out of the blade sweep has less energy because it has been slowed. The abrupt change in speed makes the wind turbulent, a phenomenon called wake. Because of wake, wind turbines in a wind farm are generally placed about three rotor diameters away from one another in the direction of the wind, so that the wake from one turbine does not interfere with the operation of the one behind it.

- Wind obstacles, such as trees, buildings, and rock formations. Any of these obstacles can reduce wind speed considerably and increase turbulence. Wind obstacles such as tall buildings cause wind shade, which can considerably reduce the speed of the wind and therefore the power output of a turbine.

- Wind shear, or differences in wind speeds at different heights. When a turbine blade is pointed straight upward, the speed of the wind hitting its tip can be, for example, 9 miles (14 kilometers) per hour, but when the blade is pointing straight downward, the speed of the wind hitting its tip can be 7 miles (11 kilometers) per hour. This difference places stress on the blades. Too much wind shear can cause the turbine to fail.

CURRENT AND POTENTIAL USES

The American Wind Energy Association predicted that in 2005 as much as 2,500 megawatts of new wind power capacity could be added in the United States, bringing the total to more than 9,000 megawatts. Worldwide, as of the end of 2003, about 39,000 megawatts of wind power were being generated, producing about 90 billion kilowatt-hours of power, enough for about nine million average American homes.

The power produced with wind energy is only a fraction of the potential. The U.S. Department of Energy says that, in theory, wind can provide the equivalent of 5,800 quadrillion British thermal units, or quads, of power each year, a number that is fifteen times the total world energy demand each year. Just a single quad has as much power as 45 million tons of coal or 172 million barrels of oil. In the United States, it is estimated that wind realistically could supply 20 percent of the nation's electricity requirements. In 2005 it was supplying about 0.4 percent. A goal is for the United States to generate 5 percent of its electricity from wind power by the year 2020.

An example of wind power in action in the United States is Spirit Lake, Iowa. At Spirit Lake, the elementary school has a 250-kilowatt wind turbine that provides 350,000 kilowatt-hours of electricity each year, more than the school needs. The rest of the power is fed into the local utility grid, earning the school $25,000 during its first five years. The school, however, is not fully dependent on the wind turbine. When the wind is not blowing, the school purchases electricity from the power company. Officials at Spirit Lake considered the system so successful that a second turbine, with a capacity of 750 kilowatts, was installed.

Commercial wind power usually is generated at wind farms rather than from single turbines. Wind farms consist of a group of turbines at the same site. The largest wind farm in the United States is the Stateline Wind Energy Center, located on the Vansycle Ridge, which runs along the Columbia River on the Washington-Oregon border. The ridge is an ideal site because of its consistent average winds of 16-18 miles (26-29 kilometers) per hour. The farm consists of 454 wind turbines, each 166 feet (51 meters) tall and at peak capacity generating 660 kilowatts of power. This wind farm provides power to about seventy thousand homes. Plans are to expand the farm so that it can produce 300 megawatts of power.

Benefits of wind turbines

Wind power has grown to be economically competitive with other forms of power. Although it costs more to generate 1 kilowatt of electricity by wind power than it does with coal- or oil-fired generators, the gap is closing. If 20 percent of a family's electricity were to come from wind power, the electric bill would be less than $2 higher per month. The cost of generating wind power has decreased 85 percent since 1980.

Wind power can be an alternative crop for farmers and ranchers. A small family farm in western Pennsylvania provides 5 percent of the power used at the University of Pennsylvania. Many farmers and ranchers are leasing their land to produce electricity. A farmer can be paid as much as $4,000 per wind turbine, and the farmer can continue to use the land for traditional farming. Wind turbines add to the local tax base. In Lamar, Colorado, wind-power generation added $32 million to the county tax base, providing money for schools and other local needs.

Wind turbines do not consume water, making them ideal for dry or drought-stricken areas. In contrast, conventional and nuclear power plants consume large amounts of water for cooling and other purposes. According to the California Energy Commission, the number of gallons of water consumed per kilowatt-hour by nuclear power plants is 0.62; by coal plants, 0.49; and by oil, 0.43. In contrast, wind-power turbines consume 0.001 gallons of water per kilowatt-hour.

Wind power is homegrown, unlike oil, which the United States and other countries have to import in large quantities from areas of the world that are often unstable. Not buying these fuels from abroad increases national security and improves the nation's balance of payments. Because wind is free, consumers are not at the mercy of frequently increasing fuel prices.

Wind power in inexhaustible and renewable, in contrast to fossil fuels, and it is clean. Wind power does not contribute to acid rain, smog, global warming, or mercury contamination. It does not release dangerous particles into the air. In 2000 the Harvard School of Public Health conducted a study on the health effects of two conventional power plants in Massachusetts. The researchers concluded that among the health effects of the plants' air pollution were 159 premature deaths, 1,710 emergency department visits, and 43,300 asthma attacks.

Wind energy is safe. Although the risk exists for industrial accidents in the construction of a wind turbine, the same can be said about the construction of any facility. The risk that the public will be harmed by a wind-power facility is nearly zero. With nuclear power the risk of catastrophe is ever present, and with fossil fuel plants, the danger from fire and explosions is high. There has been only one case of a person's being killed by a wind turbine: A skydiver sailed off course and fell into the rotating blades of a turbine.

Wind power has many uses. Small turbines can power schools, businesses, campuses, homes, farms, and ranches. They can be used in remote locations for telecommunications, ice making, and water pumping, eliminating the need for remote communities to run smoky and noisy diesel-powered generators. Turbines could benefit native communities in small, poorer nations.

Drawbacks of wind turbines

Wind turbines can be noisy, and engineers are working on ways to quiet the noise. The best method has been to reduce the thickness of the trailing edges of blades. Noise also has been reduced by placing turbines in an upwind rather than a downwind position. The wind hits the blades first, then the tower, rather than the other way around, eliminating the thumping sound that downwind designs make as the blade passes the wind shadow cast by the tower.

Wind turbine blades can cause shadow flicker as the blades rotate in the path of the sun's rays. The flickering of light and dark can be a minor annoyance for local residents when the sun is low in the sky. Most turbines are set back far enough away from homes and businesses so that shadow flicker is not a concern.

Wind farms require a fair amount of land, about 24 hectares (60 acres) per megawatt. However, the turbines themselves plus service roads occupy only about 1 hectare (3 acres) of the 24 hectares. Once the turbines have been built, farmers and ranchers can continue to use the land under them for traditional purposes. Land is difficult to find near larger cities. One solution to this problem is to place wind turbines in shallow waters offshore where possible.

Wind turbines are visible, contributing to visual or horizon pollution. Placing some wind turbines offshore can help lessen this problem. Some people consider wind turbines sleek and attractive, embodying a forward-looking concern for the environment. Wind turbines are no more visible than ski resorts, water towers, and junkyards.

The wind is intermittent, meaning that wind power has to be supplemented by other forms of power. Wind-power generation poses additional challenges for power-grid managers, who have to ensure that enough power is available to meet peak demand at all times, even when the wind is not blowing.

Not all areas of the United States, or any country, are suitable for wind-power generation. Wind towers and rotors can interfere with radar, posing a potential hazard for air travelers. They can also interfere with television and radio transmission, particularly if they are in the line of sight between the signal source and the receiver. Finally, wind turbines can be a hazard to birds, which sometimes fly into the rotors.

Environmental impact of wind turbines

The use of wind power benefits the environment, because this form of energy is clean and it does not consume water. It has been estimated that in 2004, existing wind power prevented the release of 10.6 million tons of carbon dioxide, 56,000 tons of sulfur dioxide, and 33,000 tons of nitrogen oxides. It also has been estimated that if only ten percent of wind potential were developed in the ten windiest U.S. states, total carbon dioxide emissions could be cut by one-third.

Wind power, however, can have harmful effects on the environment. Some environmentalists are concerned about soil erosion, particularly in desert regions, where a thin, fragile layer of topsoil would be disturbed in the construction of turbines, and in the eastern United States, where turbines would be built on mountain ridgelines. Good engineering practices could lessen these effects.

Another potential problem is the effects of wind farms on bird life. Although birds and bats sometimes fly into wind-turbine blades and are killed, this problem is site specific and has been exaggerated. In a study in California researchers concluded that in a total of ten thousand bird deaths, 5,500 birds were killed by flying into buildings and windows and that motor vehicles caused seven hundred deaths. Cats caused one thousand bird deaths.

Wind turbines, in contrast, accounted for less than one in ten thousand bird deaths. Environmentalists are also concerned that wind farms with their service roads and transmission lines may break up the habitat of birds and other wildlife.

Economic impact of wind turbines

The chief economic impact of wind power is that the fuel is free, so it does not have to be mined, transported, stored, and purchased by utility companies. In the early 1980s, when the first large wind turbines were being installed, the electricity they generated cost about thirty cents per kilowatt-hour. At that time, wind power was not competitive with other forms of power because it was just getting its start at a commercial level.

As the scale of wind operations grew and the technologies used to exploit wind energy improved, wind power in the early twenty-first century cost about four to six cents per kilowatt-hour, making it competitive with traditional power sources. The cost of wind power tends to be higher in the eastern United States, where wind speeds are lower, wind farms are smaller, and the cost of construction is higher because most wind turbines are constructed on elevated ridgelines. The cost tends to be lower in the Great Plains, where wind speeds are higher, wind farms are larger, and the cost of construction is lower because of the flat terrain. To put the figure of four to six cents per kilowatt-hour in perspective, the cost of electricity per kilowatt-hour in some U.S. states in 2000, according to the Energy Information Administration, was as follows:

Hawaii, 14 cents

New York, 11.2 cents

Connecticut, 9.5 cents

California, 8.4 cents

Florida, 6.9 cents

Illinois, 6.6 cents

Colorado, 6.0 cents

Nebraska, 5.3 cents

Kentucky, 4.1 cents

Wind power provides jobs. Every megawatt of wind power provides about 4.8 job-years of employment. Wind power also provides exports. It is estimated that by the mid-2010s, 75,000 megawatts of new wind power will be installed worldwide at a cost of $75 billion. Countries with the industrial capacity to build wind turbines, such as the United States, could capture a share of that growing market, providing employment for thousands of people. Many farmers and ranchers earn money by leasing their land to wind-power companies. They receive as much as $3,000 to $4,000 per year for each wind turbine. Wind farms increase local tax bases, providing funds that counties can use to improve schools and providing other services to residents.

Wind power does not have the hidden costs of other energy sources. Hidden costs are those that society has to pay but that are not reflected in the price of the resource. Such costs include transportation and storage with their risk of causing polluting accidents, air and water pollution, and the health effects of pollution.

Societal impact of wind turbines

The effects of wind power on society are difficult to measure. Because the fuel is free, use of wind power would release billions of dollars that are currently spent mining, transporting, storing, and burning fossil fuels. However, the price of land near wind turbines often decreases, which is a concern to local land owners.

Most conventional power-generating plants, and even some alternative energy plants such as hydroelectric dams, are large facilities that often alter the course of rivers and other natural landscapes. Nuclear power and coal-fired generating plants are considered necessary, but they pose health and safety dangers particularly related to smoke and other emissions. Wind power, in contrast, is perhaps the most harmless form of power available. It consumes no fossil fuels or water, it poses no health risk and only the smallest safety risk, and as the technology develops, it is likely to be relatively inexpensive, especially as the cost of fossil fuels rises. Wind energy would provide the United States, or any nation, with at least some measure of energy independence, making the nation less reliant on the energy sources that come from other parts of the world. In many communities, using wind energy would bring power generation closer to home, so that cities, counties, and states would be responsible for their own power needs. Being responsible for their own power may contribute to a greater sense of community among local residents.

ISSUES, CHALLENGES, AND OBSTACLES

Three principal issues surrounding wind power continue to be discussed by policy makers and legislators: the renewables portfolio standard, the production tax credit, and net metering.

The renewables portfolio standard, or RPS, refers to proposals for laws that would require electric utility companies to provide a portion of the electricity from renewable sources such as wind power. The company could either produce the energy itself, or it could buy the energy from another company. Rather than buying the electricity, the company could also buy credits, which it could then trade or sell to other utility companies. In this way, company A might not provide any electricity at all from renewable sources, but company B, which bought A's credit, might provide twice as much as it otherwise would have. Thus, the purpose of the RPS is not to force any single company to provide energy from renewable sources but to force the industry as a whole to provide such electricity. Twelve states have an RPS in place, and various proposals have been made to enact RPS laws at the federal level.

A second issue is the production tax credit. As a way to encourage the development of wind power, the government gives wind energy producers a 1.8-cent tax credit for every kilowatt-hour they produce. This money can be subtracted directly from the company's income tax bill, making it less expensive for the company to produce energy and therefore making the energy less expensive to consumers. In this respect, the wind industry is no different from other energy industries, all of which receive help from the tax code so that they can keep down costs to consumers. The tax credit was enacted in 1992. In 2004 President George W. Bush (1946–) signed a two-year extension to expire at the end of 2005. The wind industry would like to see the tax credit extended beyond that date so that the industry can continue making investments in wind-power plants.

A third issue is called net metering or sometimes net billing. This term refers to laws that permit citizens with wind turbines to allow their electric meter to run backward when they are supplying excess power to the electric grid. For example, a rancher has a wind turbine that generates 200 kilowatt-hours of electricity. During the day, the turbine provides much of the electricity needed to run the ranch, but when the wind is not blowing, the rancher has to buy supplemental power from the utility company. At night the turbine generates excess electricity that the rancher can sell to the local utility company.

Under net metering laws, each excess kilowatt-hour the rancher supplies would offset each kilowatt-hour he buys from the utility, lowering the ranch's electric bill each month. Some utility companies argue that this practice is unfair, because they say they are being forced to buy power from the rancher at high retail rates rather than at the low wholesale rates at which they usually buy power. The wind-power industry has successfully argued in thirty-four states that the rancher and the utility are swapping power and that this is a standard practice among utility companies. Meanwhile, other states are considering enacting net metering laws.

■ ■ ■

For More Information

Books

Burton, Tony, David Sharpe, Nick Jenkins, and Ervin Bossanyi. *Wind Energy Handbook.* New York: Wiley, 2001.

Manwell, J. F., J. G. McGowan, and A. L. Rogers. *Wind Energy Explained.* New York: Wiley, 2002.

National Renewable Energy Laboratory, U.S. Department of Energy. *Wind Energy Information Guide.* Honolulu, HI: University Press of the Pacific, 2005.

Periodicals

Linde, Paul. "Windmills: From Jiddah to Yorkshire." Saudi Aramco World (January/February 1980). This article can also be found online at http://www.saudiaramcoworld.com/issue/198001/windmills-from.jiddah.to.yorkshire.htm.

Web sites

"Energy Efficiency and Renewable Energy: Wind." U.S. Department of Energy. http://www.eere.energy.gov/RE/wind.html (accessed on July 25, 2005).

"Guided Tour on Wind Energy." Danish Wind Industry Association. http://www.windpower.org/en/tour.htm (accessed on July 25, 2005).

"Wind Energy Tutorial." American Wind Energy Association. http://www.awea.org/faq/index.html (accessed on July 25, 2005).

Energy Conservation and Efficiency

INTRODUCTION

While scientists and engineers search for alternatives to fossil fuels that are clean, abundant, safe, and inexpensive, other important alternatives are available to businesses, governments, and other energy consumers: finding ways to reduce energy use and using energy more wisely and efficiently. For the foreseeable future, solar power, wind energy, and other alternatives are likely to function mainly as supplements to fossil fuels. That is, they can meet some percentage of the world's energy needs, but the potential of these alternatives in the early 2000s is limited by cost, environmental considerations, and even simple geography. Wind power, for example, can become a major power source only in those parts of the world that have sufficient wind.

In the short term, the world will continue to rely on fossil fuels. One way to stretch the supply of fossil fuels—while at the same time reducing the pollution caused by mining, transporting, and burning them—is to burn less of them. The cost of fossil fuels is likely to increase as reserves diminish and it becomes increasingly expensive to mine or drill for less-readily available supplies. However, energy consumers can reduce their dependence on fossil fuels and their energy bills by finding new ways to use less energy. Among the best ways to accomplish these goals are increasing energy efficiency and energy conservation. The first includes redesigning vehicles, buildings, appliances, and the like—both by building them with materials that require less energy to produce and by designing them in such a way that they require less energy

Words to Know

Carbon sequestration Storing the carbon emissions produced by coal-burning power plants so that pollutants are not released in the atmosphere.

Climate-responsive building A building, or the process of constructing a building, using materials and techniques that take advantage of natural conditions to heat, cool, and light the building.

Drag coefficient A measurement of the drag produced when an object such as a car pushes its way through the air.

Green building Any building constructed with materials that require less energy to produce and that save energy during the building's operation.

Hybrid vehicle Any vehicle that is powered in a combination of two ways; usually refers to vehicles powered by an internal combustion engine and an electric motor.

Lumen A measure of the amount of light, defined as the amount of light produced by one candle.

Sick building syndrome The tendency of buildings that are poorly ventilated, lighted, and humidified, and that are made with certain synthetic materials to cause the occupants to feel ill.

Thermal mass The measure of the amount of heat a substance can hold.

Trombé wall An exterior wall that conserves energy by trapping heat between glazing and a thermal mass, then venting it into the living area.

while *in* use. The second includes the many ways in which the average person can make lifestyle choices that conserve energy, such as drying clothes on a clothesline rather than using a dryer; eating less meat; setting thermostats lower in the winter and higher in the summer; maintaining water heaters at lower settings; carpooling, using public transportation such as subways or buses, walking, or biking to work or school; purchasing smaller, more energy efficient vehicles rather than larger vehicles like SUVs; and choosing to replace incandescent light bulbs with compact fluorescent light bulbs. Some experts argue that energy conservation among consumers is a cheaper and more environmentally sensitive option to increased energy production from either fossil fuels or alternative sources.

Conserving oil and gas

Scientists and energy officials agree that the need for conservation and greater fuel efficiency is pressing, although they debate just how urgent it is. In the 1990s the Intergovernmental Panel on Climate Change (IPCC) conducted investigations that led in 1997 to the Kyoto Protocol, a worldwide plan designed to reduce fossil-fuel consumption, with the goal of reducing global warming. At that time the IPCC estimated that the amount of oil remaining in the ground was about 5,000 to 18,000 billion barrels. The panel also estimated that world production of oil and gas would begin falling in about 2050. At that point the cost of oil and gas would become painfully high. Meanwhile, according to the World Energy Council, the world consumes over 71 million barrels (one barrel equals 42 gallons) of oil and natural gas per day.

In 2003 a team of geologists from the University of Uppsala, Sweden, presented findings that differed from those of the IPCC. The good news, according to the Swedish scientists, is that global warming, caused in part by pollutants emitted from vehicles, will never reach disastrous proportions. The bad news, however, is that global warming may be less of a threat than previously thought because the amount of fossil fuels remaining is dangerously low and the world will run out of these fuels before global warming becomes a critical problem. This team of scientists believes that the remaining supply of oil is only about 3,500 billion barrels and that production will begin to fall in about 2010 rather than 2050. Furthermore, about 80 percent of the known oil and gas reserves are in regions of the world that are politically unstable, so reserves could be sharply reduced or even cut off entirely as a result of political unrest.

In general, energy experts fall into two camps, the optimists and the pessimists. The pessimists believe many countries have exaggerated their figures about proven oil and gas reserves and that all the world's major oil and gas discoveries have already been made. Thus, the pessimists believe that the world is faced with declining oil and gas supplies. In this case, energy conservation and energy efficiency are necessary because world supplies will not support the use of energy at current levels for very long. The optimists, on the other hand, believe that technological advances will lead to the discovery of more oil and gas and, more importantly, enable engineers to tap that oil and gas in ways that were thought impossible in past years. In this case, energy conservation and energy efficiency are necessary because, with the discovery of more fossil fuels, the environmental impact of using them continues to grow.

Conserving coal

Coal reserves are more abundant than oil and gas. Many experts say that the amount of coal reserves in the world—just over a trillion metric tons—is enough to last for about 200 years. The primary issue with coal, however, is that it is dirtier than oil,

Carbon Sequestration

Some scientists argue that the key to using coal without emitting huge amounts of carbon dioxide into the atmosphere is a process called carbon sequestration. Carbon sequestration refers to several methods of removing or slowing the concentration of ("sequestering") carbon dioxide in the atmosphere. According to the U.S. Department of Energy (DOE), natural sequestration occurs in various ways, including the absorption and storage of carbon by vegetation, soils, and the oceans in carbon "sinks." The DOE, many environmental groups, and some power companies support enhancing natural sequestration with methods such as the reforestation of agricultural or urban areas and restoration of wetlands, though research is needed in order to create larger, longer-lasting carbon pools in various ecosystems. Another method in development includes the capture and injection of carbon dioxide at deep sea level, though the long-term effects of injecting carbon dioxide into the oceans are unknown. In addition, several countries, including the United States, China, and England, are funding research into the capture and storage of carbon in underground or undersea geologic formations such as depleted crude oil and natural gas reservoirs, unmineable coal seams, and deep saline reservoirs. The DOE states that not only does carbon storage in depleted oil reservoirs reduce carbon dioxide levels, the pressure created can force out additional oil.

Proposals have been made in England, the United States, and other countries to rely more on carbon sequestration. England has abundant coal reserves that could provide a significant percentage of the nation's energy needs for many decades, if ways can be found to deal with the carbon dioxide. Some British experts believe that there is enough space under the North Sea to store the United Kingdom's carbon emissions for a century.

contributing significantly to the emission of carbon dioxide, the chief pollutant behind global warming. While the amount of carbon dioxide produced by burning coal differs with the type and quality of the coal, the U.S. Department of Energy provides this analysis: When coal is burned, the chief element that provides heat is carbon. During the combustion process, one pound of carbon combines with 2.667 pounds of oxygen to produce 3.667 pounds of carbon dioxide. Thus, if the carbon content of a particular grade of coal is, say, 78 percent, and burning a pound of it produces about 14,000 British thermal units (BTUs) of heat, then producing 1 million BTUs of heat releases about 204.3 pounds of carbon dioxide into the atmosphere. This figure is about twice that of natural gas and about 50 percent more than that of oil, according to the U.S. Department of Energy.

Currently, coal is used almost exclusively in the production of electricity, while oil, after it is refined into gasoline, is the primary fuel source for cars and trucks. It is possible to design and build cars and trucks that are powered by electricity, but doing so would increase demand for electricity. This increased demand would require the combustion of increasing amounts of coal, which in turn would lead to the emission of more carbon dioxide.

Conventional energy sources can be conserved in various ways by individuals, for example consumers can make conscious choices to meet rather than exceed their needs in terms of the size of their homes and automobiles. Consumers can buy smaller homes to decrease home square footage, cutting the overall energy consumed by heating, cooling, and lighting. People can decrease dependence on automobiles by utilizing public transportation, carpooling, walking, or biking. Even with minimal changes to everyday life, consumers can take steps to reduce their energy consumption by making minor improvements in their homes, such as upgrading old inefficient heating systems and installing storm windows; relying on more energy-efficient lighting and appliances; and by changing their driving habits. There are also ways in which consumers can reduce energy use by requiring the housing and automotive industries to construct climate-responsive buildings; use "green" building materials; and design and build "hybrid" vehicles that use less gasoline. Any of these can significantly cut an energy consumer's bills, reduce pollution, and help stretch the world's energy supplies.

CLIMATE-RESPONSIVE BUILDINGS

Climate-responsive building is sometimes called *green building* or *sustainable building*. In a broad sense, each of these terms refers to the same philosophy of building design and construction. This philosophy emphasizes the construction of buildings that use resources efficiently, both during their construction and once completed. Another goal is to minimize the impact of the building on the surrounding natural environment.

In this chapter the term climate-responsive building will emphasize issues pertaining to the siting (the placement), design, and layout of a building in order to take advantage of local weather conditions to reduce energy-use during the building's operation. The term green building will be used to emphasize the use of alternative construction materials that reduce energy demands. *Sustainability* is a more general term that refers to any technique,

A carpool lane sign on a California freeway near Los Angeles. © *Joseph Sohm; ChromoSohm Inc./Corbis.*

whether applied to construction or to such activities as agriculture, that enables the human community to "sustain" the natural environment for the future by using building materials and sources of energy that are renewable. These terms, though, all overlap. The design of a climate-responsive building emphasizes, in part, the use of green-building materials, and green-building practices are likely to be, at least in part, climate-responsive. The goal of both is sustainable building design.

Climate-responsive history

The history of climate-responsive buildings dates back at least to the ancient Greeks. Around 500 BCE the Greeks in many areas of the country were running out of firewood. To heat their homes, they began positioning them in a way that would take advantage of the sun's rays and provide passive solar heating. Even the philosophers Socrates and Aristotle used their influence to call for construction that took advantage of solar heat during the winter by facing transparent mica windows toward the sun. (Mica refers to a number of transparent silicates that easily separate into thin

sheets.) The Greeks also began to use dark floors and other building materials that absorbed heat during the day so that buildings
would stay warmer at night. They began to use window shutters to
trap the day's heat, and they built structures in clusters so that each
building would get some protection from cold winds.

Later the ancient Romans used similar building techniques.
Moreover, the Romans were the first civilization to use glass greenhouses not only for growing plants and vegetables but also to trap
heat. The Romans built bathhouses that took advantage of the sun,
and whole cities were laid out to provide each resident with access
to the sun—access that was protected by law. In the American
Southwest, the Anasazi Indians, in a similar way, constructed
villages that took into account the changing angles of the sun
throughout the year.

In more modern times scientists and engineers developed new
climate-responsive building techniques. In eighteenth-century
Switzerland, physicist and geologist Horace-Benedict de Saussure
(1740–1799) designed the first solar water heater. It consisted of a

The house in the background
uses solar panels on the roof
to gather energy to recharge
batteries stored in the
basement to supply power for
energy needs. The
greenhouse is used for
plants and to heat the house
through vents opening into
the upper floor windows.
©Michael Maloney/San
Francisco Chronicle/Corbis.

The new Caltrans District 7 Headquarters, located in downtown Los Angeles, can harness the energy of the sun to create electricity with a second layer of vision glass panels with special photovoltaic components located on the southern exposure of the structure. There is a special, second metal skin that can open and close to maintain indoor comfort and filter air. © Ted Soqui/Corbis.

wooden box with a black base and a glass top. The water in the box could reach a temperature of 190°F (88°C). Other scientists focused on other ways to exploit solar energy in building construction or for commercial purposes. In 1878, for example, solar energy was focused to power a steam-operated printing press in France. However, much of modern construction after that point paid very little attention to climate-responsive building. Instead, humans focused on developing artificial means of heating and cooling using fossil fuels.

In the twenty-first century, architects and design engineers have rediscovered some of these techniques. Rather than simply putting buildings anywhere and relying on fossil fuels to heat, cool, ventilate, and light them, these designers are paying more attention to local climatic conditions to make buildings far more energy-efficient. They are learning to see buildings not just as collections of steel, glass, wood, and other materials, but as systems that interact with their natural environment. By paying attention to that environment, buildings can consume less energy while still providing for the comfort of their occupants.

The need for climate-responsive buildings

Use of energy in commercial buildings is huge, so one place to start with energy conservation and efficiency is to design and construct such buildings with the principles of climate responsiveness in mind. Energy use within commercial buildings in the United States is actually higher than within the sectors of industry and transportation. And consumption of electricity within buildings doubled in the 1980s and 1990s and was expected to increase another 150 percent by 2030. As of the late twentieth century, 66 percent of the electricity used in the United States was that in commercial buildings.

In addition, buildings produce a considerable amount of carbon emissions. Buildings are responsible for 35 percent of all U.S. carbon emissions. On-site burning of fossil fuels accounts for 11.3 percent, while electricity usage accounts for 23.7 percent. Buildings also produce 47 percent of U.S. sulfur dioxide- and 22 percent of nitrogen oxide-emissions. Climate-responsive buildings can cut both this energy consumption and the greenhouse gas emissions. Such buildings can also contribute to a more healthful working climate for the building occupants.

Climate-responsive building techniques

Some of the most common climate-responsive building techniques include the following:

1. Available solar energy can be used for heating and lighting. This would include daylighting, or using natural sunlight to provide for lighting needs; solar ventilation preheating, which makes use of greenhouses, atriums, and solar buffer spaces to provide some of the building's heat; solar water heating; and photovoltaics, or the use of photovoltaic cells to provide electricity. Using daylighting and solar energy for heating and lighting requires intelligent placement of the building relative to the sun. Solar water heating and photovoltaic features can be built right into the skin and roof of the building, as well as into skylights, shingles, roofing tiles, glass walls, and even ornamental features. In fact, buildings that are constructed with built-in photovoltaics can even become net energy producers, creating surplus power that can be sold to the local energy grid or traded for power the building needs during periods when the sun does not shine.

2. Controllable shading can prevent overheating and glare. In hot-weather climates coatings can be placed on windows to block heat from entering the building while still allowing light to enter.

3. Using external wind pressure and solar radiation can power ventilation systems, serving as a supplement to fan-powered ventilation systems.

4. Using thermal mass and shading to help control internal temperatures reduces the demand for artificial heating and cooling.

5. In private homes, some builders construct Trombé walls, named after French inventor Felix Trombé (1906–1985), who conceived the design in 1964. Trombé walls are built facing the sun from materials such as stone, adobe, concrete, or even water tanks—any material that has high thermal mass (an ability to store and give off energy). The walls also have an air space, insulated glazing, and vents. As sunlight passes through the glazing and strikes the wall, the wall absorbs heat, in turn heating the air between the wall and the glazing. This warmer air then rises and is channeled through the vents into the home; cooler air from the home, which sinks, flows through vents at the bottom of the interior walls and into the air space. Heat can be retained on cloudy days by placing insulation between the air space and the thermal mass.

While incorporating energy-saving features into a building's design is beneficial, modern architects who design climate-responsive buildings make it clear that to derive the maximum possible benefit, it is important to take a "whole building approach," seeing a building not just as a collection of parts but as a living, breathing system. Further, architects point out that what works in one locale or part of the country might not work in another. A major concern in Minneapolis, Minnesota, is heating a building in winter, while residents of Phoenix, Arizona, are more concerned about cooling, especially in the summer. In the Midwest and Deep South, expelling humidity is a major concern, while in the dry air of the Rocky Mountain region, the concern is just the opposite. Architects and designers take these differing conditions and needs into account, then by integrating solar, wind, thermal mass, and other features, they can create designs that cut energy consumption significantly.

Proving this is a pair of buildings in San Diego, California. The Ridgehaven Building, a commercial office building, is located next

Sick Buildings

In addition to providing energy savings, climate responsive (and green) buildings have an additional benefit: They tend to be more healthful for the occupants, including workers in a commercial building or students in a school building. Sometimes, a building can be the site of specific illness, such as Legionnaires' disease, an illness caused by the *legionella* bacteria, which is thought to be spread through cooling systems.

Buildings, though, often suffer from what is called "sick building syndrome." This syndrome became more apparent after the rise in energy costs in the 1970s, when people started to become more aware of air leaks in buildings and sealed them to reduce wasted energy. While sealing the air leaks saved energy, it also trapped toxins and stale air inside, giving rise to a host of physical problems for the occupants, including eye, nose, and throat irritation; dryness of the skin, throat, and nose; breathing difficulties; headaches; fatigue; and even rashes.

According to the World Health Organization (WHO), sick buildings typically have forced-air ventilation, are constructed with light-weight materials, have indoor surfaces covered with fabrics, especially carpet, and are airtight. These features create uncomfortable temperatures, humidity levels that are too low, noise, and reliance on artificial lighting—especially fluorescent lighting that can "flicker" and cause headaches. They also trap molds, spores, dust mites, and other microorganisms. Some equipment such as photocopiers and printers may have toxic solvents in their toners, while carpeting and adhesives release toxic vapors such as formaldehyde.

Some experts, such as those at the Renewable Energy Policy Project (REPP), argue that sick building syndrome has a distinct economic cost and that climate-responsive buildings can lessen those costs. According to the REPP, such features as daylighting and natural ventilation can reduce employee sick days, boost the achievement of school students, and even increase sales in retail outlets. The REPP says that a ten-percent improvement in the productivity of employees can actually pay back the entire cost of a building over a ten-year period.

door to a nearly identical building of the same size. The Ridgehaven Building was built using climate-responsive techniques and with green materials; the neighboring building was constructed using traditional techniques and materials. The Ridgehaven Building uses 65 percent less energy than the neighboring building, saving the building's owners $70,000 a year in utility bills.

GREEN BUILDING MATERIALS

Closely related to climate responsiveness is the concept of using "green" building materials. "Green" is a word that is used in connection with environmentally sustainable building materials and practices. It does not refer to the actual color of the materials. Rather, because green is the predominant color of the natural world, the word has become a figure of speech to refer to any environmentally sound practice that reduces the impact of human activity on the natural environment.

The need for green building materials

Many green building practices have goals other than energy efficiency. For example, using products made out of natural materials can reduce the level of toxins and other harmful substances in a building. These substances are emitted by such materials as synthetic carpets, adhesives (e.g., glue used to bind two elements together), and fiberglass insulation. Substituting materials made from natural products (like insulation made from recycled paper) can contribute to the health, and therefore productivity, of employees working in a commercial building. Other green building practices are designed to reduce water consumption, for example toilets that use less water and landscaping that does not require large amounts of water. Still other practices are designed to minimize waste. One simple technique is to design buildings with dimensions that use entire 4- by 8-foot (1.2- by 2.40-meter) sheets of particleboard rather than creating large amounts of scrap. Also, using other green building materials that are made from recycled materials. Roof shingles, for example, can be made from recycled vinyl and sawdust.

Many green building practices, however, have energy efficiency and conservation as their primary goal. Many green construction materials save energy not only in the day-to-day operation of the building but also in its construction, because producing and transporting the materials are less energy-intensive activities. Furthermore, some green building materials are more durable than their traditional counterparts. This represents a form of energy conservation because the structure will last longer. A good example is cement composite house siding. Used more and more in place of wood, the cement composite can last fifty years or more with virtually no maintenance, primarily because it is not only tough but the color is mixed into the composite rather than applied on the surface, so it does not have to be painted. Though the initial production of cement is more costly in terms of carbon dioxide emissions than wood, the energy-efficiency of a building made with cement composite may save more carbon dioxide emissions over the lifetime of the building than were used making the cement.

Common green materials

Below are some twenty-first century green building materials. Most of these are more practical for houses than they are for commercial construction such as office buildings. Nonetheless, the impact of using these materials in large numbers of homes could be considerable.

1. Adobe: Adobe is one of the world's oldest building materials. Essentially, adobe is nothing more than earth that has been mixed with water and shaped into bricks. Sometimes chopped straw is added to give the adobe additional strength. Adobe is most durable when the content of the earth is about 15 to 30 percent clay, which binds the material together. The rest is sand or aggregate (small bits of rock). While adobe is commonly used in the southwestern United States, it can be used in most areas of the country. The chief advantage of adobe is that it provides good thermal mass, meaning that it absorbs heat during the day, then slowly releases the heat during the cooler nighttime. Some homeowners use adobe because the walls absorb heat during the day, then transfer the heat to the main portion of the house at night. The chief disadvantages of adobe are that it is structurally weak and is not a good insulator. Thus, adobe homes are often built very thick and may include a layer of insulation.
    A variation of adobe is called cast earth, which consists of blocks made of a mixture of earth and plaster of Paris. The plaster gives the blocks greater strength, so the amount of clay is unimportant. Cast earth has a strong aesthetic appeal to some builders because of its stone-like appearance.

2. Cob: Cob, which was commonly used in nineteenth-century England, is similar to adobe, but it has a much higher straw content. Because of the additional straw, it works better as an insulator than adobe does, though cob is often much thinner than adobe construction it is also becomes rather brittle over time. Another difference is that while adobe is typically fashioned into bricks, cob is applied in a more freeform manner, similar to plaster. This can give structures a more artistic look. A variant of cob is called light straw. With light straw the primary component is the straw itself, which is bound together with an adobe-like mixture. Light straw has even higher value as an insulator.

It is more fragile, though, so it has to be used with a timber frame to bear loads.

3. Rammed earth: Rammed earth is another very old construction technique. Much of the Great Wall of China consists of rammed earth. Rammed earth construction again is similar to adobe in that it makes use of local materials. Rather than shaping the earth into bricks (as with adobe) or applying it like plaster (as with cob), rammed earth refers simply to the process of compressing large amounts of earth into thick walls. Often, a stabilizing ingredient, such as cement or even asphalt, is added to the earth to make it more stable and durable. Wooden or metal forms are used to give shape to the walls, in much the same way they are used in pouring a concrete foundation. Like adobe, rammed earth provides a great deal of thermal mass but is not a good insulator. Another disadvantage is that rammed earth is very labor-intensive, usually requiring considerable use of heavy equipment.

4. Earth bags: Some builders are experimenting with bags of earth, similar to the sandbags that are used for flood control. Builders fill the bags with adobe material or use crushed volcanic rock, which provides greater insulation. The bags are laid in courses, similar to brick, then covered with some sort of plaster-like substance. Many builders are turning to a covering called papercrete, which consists of shredded recycled paper mixed with cement.

5. Straw bales: Bales of straw are one of the most common green materials used in home construction, primarily as an insulator. The home is constructed using traditional framing methods. The chief difference is that much more space is left between the interior and exterior walls. This space is filled with bales of straw rather than fiberglass insulation, which is made from petroleum and therefore depletes petroleum reserves. Not only is the straw a good insulator, but many homeowners like the thick walls and deep windowsills that result from straw bale construction. Straw bale homes are also quiet, because the straw acts as a sound insulator. The chief disadvantage is that great care must be taken to prevent water from getting into the walls and to prevent the buildup of condensation, because moisture can cause the straw to rot.

Thermal Mass

Energy experts always refer to thermal mass, which measures not the flow of heat but the amount of heat that a substance can hold. Thermal mass is important primarily in areas where there are wide temperature swings throughout the 24-hour day, such as the southwestern United States and parts of the Rocky Mountain region. During the day, as outside temperatures rise, the temperature of the outside of a house is higher than that of the inside. Thus, following the laws of thermodynamics, the heat flows from outside to the cooler inside. During the night, when temperatures tend to fall dramatically (primarily because in these regions the air is drier, so there is no blanket of humidity to trap the day's heat), the heat flow reverses. Heat now flows from the warmer inside of the house to the cooler outside. But thermal mass is always responsible for a time lag. It might take up to eight hours for heat to move from outside to inside in the daytime—but by that time, the sun has set and the heat flow has stalled and starts to reverse. Likewise, it might take up to eight hours for heat to move from outside to inside, but by that time the sun is rising, so once again the heat flow is reversed. The key point is that thermal mass, as in an adobe home, helps to keep the inside temperature relatively constant, so that it changes far less than the outside temperature. A building with a great deal of thermal mass "holds" the heat rather than transferring it.

Thermal mass is a much less important consideration in areas of the country where the temperature does not swing as dramatically. In the north, for example, the daytime high temperature in the winter is almost always lower than the indoor temperature; similarly, in the summer the nighttime low temperature is very often higher—or nearly so—than a comfortable indoor temperature. Because the heat flow does not reverse itself under these conditions, thermal mass is less important.

In addition to these common green building materials, builders have experimented with many other types of materials, all with a view to reducing energy consumption and recycling materials that would otherwise find their way into landfills. Some builders, for example, build walls out of recycled tires. They fill the tires with earth, stack them, then plaster over the walls so that the tires do not show. This type of construction, in combination with other methods such as passive solar design and bermed (mounded or piled up) earth on the north side of the house, contributes to very low energy bills for the homeowner.

Embodied energy

In addition to focusing on the energy savings of new climate responsive buildings and use of green building techniques, the "embodied energy" of existing structures must be taken into account. Embodied energy is basically all of the energy (beyond that of the operating costs such as heating and lighting of the building itself) used during a building's life cycle. This would include things such as recycling or removing previous structures; harvesting wood or other resources used in the building; manufacturing other materials used in the building; and transporting materials to the site. In many cases, older buildings contain large amounts of embodied energy, so it consumes less additional energy and is more environmentally friendly to upgrade or restore the older building than to demolish and rebuild, even if green materials are used in the new construction.

When a building is demolished, all of the non-renewable energy used to create the original building is lost and more must be used to rebuild. There are several reasons why remodeling older buildings for efficiency may be a better environmental choice than destruction. The demolition and removal of materials can take up huge amounts of landfill space. Reusing old materials prevents the destruction of more trees, saves the energy used to transport them to mills and create new construction materials, and keeps more green space from development. And, since the energy used to create the original structure has already created pollution, especially with materials such as concrete, which is responsible for large amounts of carbon dioxide during production, tearing down the old structure means that all of the pollution created in building the original structure will be followed by more pollution caused in the creation of a new building.

LIGHTING

Energy experts estimate that up to one-quarter of a typical homeowner's energy bill is for artificial lighting. While climate-responsive building techniques can help lower energy use by situating homes and buildings in a way that takes more advantage of natural light, doing so may not be possible for existing buildings, which have to continue to rely on artificial lighting. Further, even the best positioning of a home to take advantage of the sun is of little use on a cloudy day or after the sun has set. Nonetheless, building occupants can take steps to conserve energy on lighting.

Did Thomas Edison "Invent" the Lightbulb?

The short answer to this question is "yes and no." In 1860 British physicist and electrician Joseph Wilson Swan (1828–1914) invented an incandescent bulb using a carbon paper filament, but the bulb did not work very well. He abandoned the pursuit for 15 years, but he returned to the problem in 1875. In 1878, a year before Edison, he demonstrated a working incandescent lightbulb with a carbonized thread as a filament. Edison receives all the credit for the invention of the incandescent lightbulb because he developed the first bulb that was commercially successful.

When it was pointed out to Edison that most of his experiments were failures, he famously commented that they were not failures but successes, for he had successfully discovered that the substances he tried did not work.

Incandescent lightbulbs

Until Thomas Edison invented the incandescent lightbulb in 1879, artificial light was produced primarily by candles and oil lamps, which were not only inefficient but also produced a fire hazard. For years during the nineteenth century, inventors experimented with ways to produce artificial light by passing electricity through some sort of filament in a vacuum. These experiments, however, repeatedly failed because the filament quickly crumbled as a result of the intense heat that made them glow. After numerous experiments testing about a thousand materials, Edison finally came up with one that worked: a carbon-based filament. His earliest lightbulbs burned for an average of about 170 hours before the filament crumbled.

Today the typical incandescent bulb—a design that has not changed much since Edison's day—lasts about from 750 to 1,000 hours, although more expensive long-lasting bulbs can last 2,500 hours. The bulb consists of a thin, frosted-glass "envelope" that houses the filament, which today is made of the element tungsten, as well as an inert gas (argon). Inert gases are used to fill the bulb for two reasons. One, the bulb cannot contain any oxygen; if it did, the intense heat of the filament would set the bulb on fire. Two, because a gas like argon is "inert," meaning that it does not combine with other elements, tungsten atoms that evaporate from

In the Limelight

The traditional lightbulb is not the only form of incandescence. Incandescent light can also be produced by a rod of lime (a highly flammable solid) surrounded by a flame fueled by oxygen and hydrogen. In the nineteenth century this type of light was the brightest form of artificial light known. Its primary use was to light stages in theaters. This is the origin of the expression "in the limelight," or being in the public's attention.

the filament bounce off the argon and most are redeposited on the filament, making the bulb last longer. The filament in a 60-watt lightbulb is about 6.5 feet (2 meters) long, but only about one one-hundredth of an inch thick. It is wound into coils so that it can fit into the bulb. Electricity is applied to the filament, exciting the atoms and producing light. A bulb eventually burns out because the tungsten in the filament evaporates and some of it deposits on the glass. In time, the filament develops a weak spot where it breaks.

Incandescent lightbulbs have a number of advantages. They are inexpensive and easy to use, and the quality of the light they produce is good. (They are so inexpensive that before the energy crises of the 1970s, some electric companies provided lightbulbs to their customers free, usually exchanging new bulbs for burned-out ones.) They can also be used with dimmer switches. But a chief disadvantage is that they are not energy-efficient. After an incandescent lightbulb has been on for a brief period of time, it becomes hot to the touch. This is because the electricity heats the filament to 4,500°F (2,500°C). In other words, most of the electrical energy going into the bulb is converted into heat rather than light. In this respect, an incandescent lightbulb is little different from an electric space heater or a toaster. This production of heat is a double disadvantage in hot-weather climates, where buildings have to be air-conditioned, because a large number of incandescent lightbulbs add to a building's interior heat, placing greater demands on the air-conditioning system. Thus, electricity is being wasted twice.

A more recent innovation is the halogen lamp. The basic technology of a halogen lamp is similar to that of the incandescent bulb. A halogen bulb uses a tungsten filament, but it is encased in an envelope made of quartz rather than glass. Further, this envelope is positioned very close to the filament, but since it is made of quartz, it does not melt. The quartz envelope is filled with gases from the halogen group, consisting of fluorine, chlorine, bromine, iodine, and astatine. What is unique about these gases is that they combine with tungsten vapor. As the tungsten of the filament evaporates, its atoms combine with the gases and then are redeposited on the filament. Thus, halogen lightbulbs last much longer than incandescent lightbulbs. Combined with a parabolic reflector, they produce a high-intensity, crisp light, making them useful for such items as car headlights, most of which are now halogen. The chief disadvantage is that they are energy wasters, for they get even hotter than incandescent bulbs, creating up to four times as much heat. Halogen lamps can be a serious fire hazard in a home, especially if they are too close to draperies or other flammable materials.

Fluorescent lightbulbs

Fluorescent lightbulbs were first invented in 1896. Today they are more commonly used in commercial buildings than homes, although many homeowners use fluorescent bulbs in basements, workshops, and laundry rooms. They tend to be less popular in the living areas of a home for three reasons. First, they often have a subtle flicker, which at best is an annoyance and at worst can cause headaches for some people. Second, the quality of the light they give off tends to be less "warm" than that emitted by incandescent bulbs, which give off more light from the red end of the light spectrum and less from the blue end, in contrast to fluorescent bulbs. For many people, fluorescent lighting has a kind of "sickly" look, although modern fluorescent light has largely overcome this problem. Third, they tend to be a bit noisy, emitting a low hum, although this disadvantage, too, has been overcome by recent technology. The chief advantage of fluorescent lighting is that it is much more energy-efficient than incandescent lighting. Further, fluorescent lightbulbs last 10-15 times longer than incandescent bulbs—often up to 10,000 hours or more.

To measure that efficiency, a distinction is made between *watts* and *lumens*. A watt is a measure of electrical usage equal to 1/746th of a horsepower, or one joule per second. (A joule is a unit of energy equal to the work done by a force of one newton acting through a distance of one meter; a newton is the amount of force needed to impart an acceleration of one meter per second per second to a mass of one kilogram.) Typically, the size of an electric lightbulb is measured in watts. Thus, found throughout a typical home are likely to be bulbs of different wattages, such as 40- or 60-watt bulbs where less light is needed and 75-, 100-, and 120-watt bulbs where more light is needed, especially for reading or similar activities.

Wattage, though, measures electrical usage. It is not a measure of the amount of light the bulb produces, although higher watt bulbs are likely to produce more light. Light output, on the other hand, is measured in lumens. Defining a lumen is much easier than defining a watt. One lumen is equal to the amount of light emitted by one candle. The 40-watt incandescent bulb made by one major manufacturer emits 475 lumens, the 60-watt bulb emits 830 lumens, and the 100-watt bulb emits 1,550 lumens.

Fluorescent bulbs produce the same number of lumens as incandescent bulbs with about one-fourth to one-sixth the amount of wattage—that is, electricity. Thus, fluorescent bulbs are far more energy-efficient than incandescent ones. They achieve this greater efficiency because they do not produce nearly as much waste heat, so per watt of electricity consumed, they produce more lumens.

Fluorescent lightbulbs are easily recognizable because rather than being shaped like bulbs, they are tubular. This sealed glass tube contains mercury and an inert gas (such as argon). The inside of the tube is coated with phosphor powder, a substance that emits light when its atoms are excited. At each end of the tube is an electrode that is wired to an electrical circuit. When the current is turned on, the voltage across the electrodes causes electrons to move from one end of the tube to the other. The energy converts the mercury from a liquid into a gas. The electrons collide with the mercury atoms, exciting them so that their electrons move to a higher energy level and higher orbit. As the electrons move back to their original orbits, they emit light.

The process, though, does not stop there. The light that is emitted is in the ultraviolet wavelength range, so it is not visible. This is where the phosphor powder coating goes to work. The photons created during the first step in the process collide with the phosphor atoms, again exciting them and causing their electrons to move to a higher energy level. Once again, when the electrons return to their normal energy level, they emit photons. These photons have less energy than the original photons; this is because some of the energy is released in the form of heat. But these lower energy photons now give off light that is visible, so-called white light that the human eye can detect. By using different combinations of phosphors, bulb manufacturers can alter the color of the light.

For many years, one of the problems with fluorescent bulbs was that it took them several seconds to light up. "Rapid start" lights have been developed to overcome this problem. In these lights a mechanism called the ballast maintains current through the electrodes. When the light is turned on, the electrode filaments heat up very quickly to ionize gas in the tube. Modern ballast mechanisms have also helped to reduce or eliminate both the flicker and noise that earlier ballasts created.

Compact fluorescent bulbs

Traditional fluorescent bulbs are long, thin tubes rather than actual "bulbs," making them unsuitable for use in floor and table lamps and even in many wall and ceiling fixtures. For this reason, they have been used primarily in special ceiling fixtures in commercial buildings, as well as in certain areas of the home. Further, they cannot be used in regular lamps or fixtures because of the nature of the plug, which consists of pairs of pins at each end rather than the metal screw portion of an incandescent lightbulb.

In the 1980s these shortcomings were corrected with the development of the compact fluorescent lightbulb (CFB). This type of bulb works in exactly the same way that a traditional fluorescent bulb does, but rather than being packaged in a long tube, the tube is smaller and folded in such a way that the bulb resembles a traditional incandescent bulb. Further, rather than pins at each end, the bulb screws into the light fixture in exactly the same way incandescent bulbs do (although occasionally some of these bulbs require special fixtures because the screw portion is a different size).

What this means is that fluorescent lighting can now be used throughout a home or other building, with the potential for enormous energy savings. The California Energy Commission estimates that a single 20-watt compact fluorescent bulb used in place of a 75-watt incandescent bulb (remember that fluorescent bulbs produce more lumens per watt than incandescent bulbs do) will save 550 kilowatt-hours of electricity over its lifetime. It takes about 500 pounds (227 kilograms) of coal to produce this much electricity, and burning this amount of coal releases about 1,300 pounds (590 kilograms) of carbon dioxide and 20 pounds of sulfur dioxide into the atmosphere. That is just one bulb. It has been estimated that if every American used CFBs, the nation could save 31.7 billion kilowatt-hours of electricity each year. A typical coal-fired power plant produces about 500 megawatts, or about 3.5 billion kilowatt-hours, of electricity per year. To generate this electricity,

Typically the amount of energy used by individual homes is measured by a meter attached to the outside of the house. © 2005 Kelly A. Quin.

it has to burn about 1.43 million tons of coal, releasing 10,000 tons of sulfur dioxide and about 3.7 million tons of carbon dioxide. Converting all home lighting to CFBs would in effect eliminate the need for roughly nine of these power plants.

CFBs have one disadvantage. While a typical incandescent bulb costs about $0.75, CFBs average about $11. The tradeoffs, though, are significant energy savings over the life of the bulb, combined with the fact that the bulb is likely to last up to ten times longer.

ENERGY EFFICIENCY AND CONSERVATION IN THE HOME

Climate responsiveness and the use of green building materials are options for new home construction. Most people, however, do not have this option because their homes were constructed years ago before these innovations were widely used. Nonetheless, homeowners can take many steps to lower their energy bill by saving energy. Some of these steps involve changes they can make to the home itself to conserve energy; others involve steps they can take to reduce personal energy use or use energy more efficiently.

Energy conservation

Experts recommend the following as ways to conserve energy in the home—many of these same steps can be taken in commercial buildings as well.

1. Phantom loads. Many electronic devices use electricity even when they are turned off. Such items as videocassette recorders, televisions, microwave ovens, and computers, as well as business machines such as copiers and faxes, all consume energy when they are not in use. A simple way to lower energy use with these devices is to plug them into a power strip, which can be turned off when the device is not being used. Another way is to unplug wall transformers (such as those used to charge a battery in a power tool or a cell phone) when they are not needed. A wall transformer, even if a tool or appliance is not plugged in, still operates and is warm to the touch. This warmth represents wasted energy.

2. Hot water. A major component of a family's energy bill is for hot water—typically about one-seventh of a home's energy bill. Hot water tanks, especially older ones, can be insulated with kits available at hardware stores. Point-of-use hot water heaters, which operate only when the hot water tap is turned on, reduce the need for a standing tank of hot water that is not being used. Most manufacturers preset the temperature on hot water heaters at 140°F (60°C), but 120°F (49°C) is sufficient for most households (and reduces the risk of scalding by water that is too hot). Lowering the thermostat temperature on a hot water heater by 10 degrees can save 3-5 percent on hot water costs. Moreover, low-flow shower heads—those that flow at a rate of 2.5 gallons (9 liters) per minute or less rather than the 4-5 gallons (15-19 liters) per minute of older shower heads—reduce the consumption of hot water, saving energy. One commonsense way to reduce hot water consumption is not to let the shower run for long periods of time while preparing to get in.

3. Heating and cooling. Thermostats can be turned down at night and when the family is away for the day. A programmable thermostat can be set to turn the heat down at night or during times when no one is at home, then warm the house up just before the family gets up in the morning or just before they are scheduled to return home at the end of the

day. Also, the style of indoor dress can be changed slightly so that indoor temperatures can be set lower in winter and higher in summer. In the summer, fans may be used to compensate for decreasing the use of air conditioning. Weather stripping can reduce heat loss around leaky doors and windows. Insulated curtains can help reduce heat loss through windows at night. Double-paned thermal windows allow the warmth of the sun to enter the home when the sun is low in the winter sky but block the sun's heat when the sun is high in the summer sky, reducing the need for air conditioning. Rooms that are not being used can be closed off and the heating in the room turned off (or the hot air duct closed). Changing filters on furnaces and having the furnace serviced each year can reduce energy consumption.

4. Insulation. Because heat rises, most heating energy is lost through a home's roof. An investment in a few hundred dollars' worth of insulation can reduce home heating (and cooling) bills by as much as 30 percent. Insulation can be installed in ceilings. Contractors can even insulate existing exterior walls by blowing insulation through small holes drilled between wall supports.

5. Landscaping. Well-placed landscaping can reduce heating and cooling bills. Deciduous trees (those with leaves) can be placed so that they block the sun, especially on the south side of a house, during the summer. The trees then lose their leaves in winter, allowing sunlight through to warm the house. Windbreaks, consisting of a row of trees or bushes, especially on the north side of a house in most areas, can block winter winds, lowering heating bills. According to Colorado State University researchers, windbreaks in some areas can reduce heating bills by as much as 25 percent.

Energy efficiency

One major way to conserve energy is to use energy more efficiently. Using compact fluorescent lightbulbs, double-paned thermal windows, and insulation conserves energy by enabling homeowners to heat, cool, and light their homes more efficiently. But another way to conserve energy is to use appliances that consume less energy.

Beginning in the 1980s the United States Congress passed several laws mandating minimum energy efficiency for appliances such as refrigerators, freezers, washers, dryers, ovens, water heat-

Energy efficient fluorescent light bulb with EPA Energy Star. © *Peter ZiminskiVisuals Unlimited. Reproduced by permission.*

ers, and pool heaters. Smaller manufacturers make many of the most efficient appliances, which tend to be more expensive. But even the major manufacturers have models that are far more energy efficient than appliances used to be. Here are some guidelines that promote energy efficiency in appliances:

- Refrigerators: Models with the freezer on top are generally more efficient than side-by-side models and those with the freezer on the bottom. Refrigerators that have to be defrosted by hand use about one-half the energy of automatic-defrost models. The most efficient refrigerators tend to be in the

16- to 20-cubic-foot range. Generally, though, it is more efficient to run one large refrigerator than two smaller ones.

• Washing machines: Many homeowners overuse the hot wash cycle. The warm and cold settings are adequate for most laundry. Energy-efficient washers automatically control the water level for the size of the load. Also, the spin cycle, in which the machine spins quickly to eliminate as much water from the clothes as possible, is faster in energy-efficient washers. Thus, more water is expelled from the clothes, and they do not have to spend as much time in the dryer. Horizontal axis machines—that is, front loaders—use far less water and soap and are much more efficient than vertical axis machines, or top loaders. The cost of running a front loader is about one-third that of running a top loader. One major manufacturer makes a washing machine that communicates with the dryer and presets it to deliver the most efficient results.

• Clothes dryers: The most energy-efficient clothes dryer is the sun and a line to hang the laundry on. In rainy or cold weather, racks for drying laundry can be used indoors, and the humidity the drying clothes add to indoor air is an added plus.

• Dishwashers: One way to boost the energy efficiency of dishwashers is, of course, not to use them as often and only for full loads. Many dishwashers have a "no-dry" cycle that saves energy; the dishes air-dry instead of being dried by heat produced by the dishwasher itself. Also, many dishwashers have water heaters so that only the water going to the dishwasher is being heated.

The guidelines for energy efficiency focus on conventional appliances, like those that can be purchased at such places as department stores. For consumers who want to achieve even greater savings on their energy bills, specialty products are available. Examples include solar-powered hot water heaters (especially heaters for smaller quantities of water, enough, for example, for one person to take a shower); solar cookers that focus the sun's rays to produce enough heat for cooking purposes or straw ovens that store the heat in the heated food to cook it; washing machines that require no electricity, relying instead on soaking and using a hand crank to wring out water; and point-of-use water heaters that activate when the hot water tap is turned on and heat just the water that is being used rather than a tank of standing water.

Solar cookers use alternative technology to generate heat from the sun. *Joyce Photographics/Photo Researchers, Inc.*

One conventional appliance that has potential for significant energy savings is the refrigerator, which on average uses about nine percent of the energy consumed in homes. Standard refrigerators and freezers use about 3,000 watt-hours per day, although it is possible to find commercial models that use just 1,500 watt-hours per day. Some manufacturers, however, build superinsulated refrigerators that use only about 750 watt-hours per day, depending on

Energy Star Ratings

Energy experts urge consumers to look for the Energy Star label when they shop for appliances. The label appears on appliances such as refrigerators, washing machines, dishwashers, water heaters and heat pumps, and even on windows. The label indicates that the energy efficiency of the appliance exceeds that required by federal regulations. Appliances that earn the Energy Star label are at least 13 percent more efficient than normal machines, but many are 15, 20, and even 110 percent more efficient. For example, Energy Star washing machines use 50 percent less electricity than those that do not have the Energy Star label.

the size and model. Smaller superinsulated refrigerators use only 200 watt-hours per day. These types of refrigerators are ideal for people who run their homes primarily on solar power. Fewer solar panels have to be added to the home to power the refrigerator.

TRANSPORTATION

Energy savings in the home and in commercial buildings makes a vital difference in the total amount of energy consumed. Still, the energy used to power cars and trucks represents a major portion of energy expended. Just in the United States, drivers consume about 360 million gallons of gasoline each day, or about 131 billion gallons of gasoline each year. If one gallon of gas, when burned, releases about 5-6 pounds (roughly 2.5 kilograms) of carbon dioxide into the atmosphere, then U.S. drivers are releasing about 2 billion pounds of carbon dioxide into the atmosphere each day. While U.S. drivers consume about 45 percent of the world's gasoline, they are not responsible for the entire problem with vehicle gasoline consumption. As of 2005, for example, the number of private cars in Beijing, the capital of the People's Republic of China, was 1.3 million, up 140 percent just since 1997. In 2005 China consumed about 252 million gallons of gasoline per day, but that figure was predicted to double to 504 million a day by 2025. Meanwhile, according to the World Bank, sixteen of the twenty most polluted cities in the world are in China, and vehicles cause most of that pollution.

Before the energy shortages of the 1970s, Americans tended not to care very much about what kind of gas mileage their cars got.

Large, "gas-guzzling" cars were the norm, and gasoline was relatively inexpensive, so little attention was paid to gas mileage. In the 1960s it was not uncommon for a family car to get as little as 10 miles (16 kilometers) per gallon or even less. Beginning in the 1970s, though, efforts were made to improve the gas mileage of cars by making them smaller and lighter and by introducing technical innovations that enabled them to burn gas more efficiently. While cars became more efficient in the following years, Americans also developed a taste for larger, heavier vehicles such as sport utility vehicles (SUVs). Thus, by the year 2000 many Americans were driving vehicles that got the same gas mileage as those that they drove 25 years earlier.

HYBRID VEHICLES

The early 2000s saw the introduction of so-called hybrid vehicles. A "hybrid" of any sort is a combination of two or more features that produces a benefit. In the case of vehicles, a hybrid combines two technologies for using energy in a way that reduces energy consumption. While conceivably any two technologies might be used in hybrid vehicles, the most common is to combine a conventional internal combustion engine with an electric motor and batteries that power the car with electricity. In the future, hybrids are likely to make use of other technologies, including hydrogen fuel cells and possibly even steam power.

Hybrid vehicles are not entirely a new concept. The moped, a motorized pedal bike, is a hybrid vehicle that combines a gasoline motor with pedal power. Locomotives are diesel fuel–electric hybrids, as are many giant trucks used for mining. Submarines, too, are hybrid vehicles using diesel-electric and in many cases nuclear-electric combinations. In 1899 German automaker, Ferdinand Porsche (1875–1952), engineered a hybrid car. The current generation of hybrid vehicles uses a combination of gasoline and electricity for power, as did Porsche's car.

The hybrid design overcomes the chief disadvantages of all-electric cars. Cars powered entirely by electricity have to be plugged in to a power source when they are not in use. These cars have limited range—generally about 100 miles or so—before the electrical power stored in the car's batteries is depleted. Moreover, the process of "refueling" is time-consuming and inconvenient. In a hybrid car the gasoline-powered engine and the batteries work with one another. Typically, an electrical motor, powered by batteries, powers the car's engine. The internal combustion engine

A multi-information display provides the driver with feedback regarding the vehicle's use of energy in the Toyota Prius. Yellow arrows indicate when the battery is in use. *Leonard Lessin/Photo Researchers, Inc.*

provides a power boost when necessary, especially when the car is accelerating. The gas-powered engine keeps the batteries charged, so the car does not have to be plugged in. On some models, when the car is idling, the internal combustion engine does not operate, so no gas is being consumed. This feature makes hybrid cars very quiet when the car is stopped at an intersection.

The components of a hybrid vehicle

A typical hybrid vehicle consists of the following components:

- Gasoline engine: A hybrid has a gasoline-powered engine similar to that found on a standard vehicle. This engine, however, is small and more fuel-efficient than the engine on a normal vehicle, boosting gas mileage and lowering emissions.
- Fuel tank: The hybrid has a tank for storing gasoline.
- Electric motor: Hybrid vehicles use sophisticated motors to provide some portion of the power the vehicle needs and to recharge the batteries.

- Generator: In some hybrids the motor acts both as a motor and as a power generator. In others a separate generator produces electrical power.

- Batteries: A battery pack stores energy produced by the motor and braking system. One major problem with electric vehicles is that gasoline is much more energy dense than batteries. That is, one gallon of gasoline contains as much energy as 1,000 pounds (454 kilograms) of batteries. The advantage of hybrids over all-electric vehicles is that the battery pack does not need to be as large because the motor is continually recharging the batteries.

- Transmission: The transmission of a hybrid is similar to that on a standard car, although some manufacturers have introduced more sophisticated transmissions that can be powered both by the electric motor and by the gas-powered engine.

Advantages of hybrid vehicles

Hybrid vehicles have at least two advantages. First, a hybrid's internal combustion engine is generally much smaller and more fuel-efficient than the engine of a standard car. This is because the engine does not do all the work. It is assisted by the batteries that supply power to the car's drive train. Generally, the internal combustion engines of standard cars are much larger than they need to be. A standard car might be capable of 200 horsepower or more, but a car generally needs only about 20 horsepower to overcome drag as the car pushes its way through air, to compensate for the friction produced by the tires and transmission, and to power such accessories as the power steering and air-conditioning. All the extra power is used primarily for sudden acceleration or to climb an uphill grade, but that extra capability is used only about one percent of the time the car is on the road. Therefore, in contrast to big, high-horsepower engines, hybrids use smaller, lightweight engines. One model's engine weighs only 124 pounds (56 kilograms), has only three cylinders (as opposed to the six or eight cylinders on many larger vehicles), and produces just 67 horsepower. By using small engines and designing them so that they operate at close to their maximum load, hybrid vehicles cut down on gas consumption.

A second advantage is that hybrid vehicles make use of what is called a regenerative braking system. Such a system is based on the laws of thermodynamics, which say that energy cannot be created or destroyed but can only change form. When a car is moving

The 2005 gas/electric Toyota Prius with Hybrid Synergy Drive offers better fuel efficiency than typical automobile engines. © *Ted Soqui/Corbis.*

down the road, it burns gasoline, releasing energy that is converted into the mechanical energy of the car's drive train. Some of the energy is lost to friction where the tires meet the surface of the road, as well as in the transmission. But much of a car's energy is lost when the brakes are applied, converting the kinetic energy of the moving car into friction, which is released in the form of heat in the car's brakes. (This explains why cars periodically need a brake job to replace the brake pads, which have been worn down by heat.) A hybrid vehicle recaptures some of this otherwise lost energy and sends it off into the car's batteries, where it is recycled to power the car. The end result is vehicles that generally get much higher gas mileage—up to 60-plus miles (97 kilometers) per gallon for some models—and that release one-tenth the amount of pollution into the atmosphere compared to standard vehicles.

Hybrid manufacturers incorporate other ways to increase the fuel efficiency of their vehicles. They recover energy and store it in the battery and allow the gasoline-powered engine to shut down when the car is idling. In addition, they use advanced aerodynamics to reduce drag. The chief way this is accomplished is by

Drag Coefficient

Engineers use the term *drag coefficient* to refer to measurements they make of the amount of drag a vehicle generates as it pushes air out of the way while it is in motion. Engineers can calculate the drag coefficient of various shapes under normal conditions. Thus, the drag coefficient of a sphere is 0.47; of a cube, 1.05; of a long cylinder, 0.82; of a short cylinder, 1.15. The most aerodynamic shape—that is, the one with the lowest drag coefficient—is the streamlined "teardrop" shape with the pointed end at the front, at 0.04. Energy-efficient vehicles cannot use a pure teardrop shape, but they can use something that approaches it by reducing the front area of the vehicle.

reducing the front area of the vehicle so that the volume of air the car has to push through is reduced.

Automakers have even found ways to reduce the drag caused by objects such as mirrors that stick out from the vehicle. Some have replaced side mirrors with small cameras. Others partially cover the rear wheels to reduce drag and also enclose parts of the undercarriage (the underside of the car) with plastic panels. The result is a very low drag coefficient, sometimes as low as 0.25. Hybrid makers often install low-rolling resistance tires. These tires are stiffer and inflated to a higher pressure than standard tires, two aspects that reduce drag by as much as half. Finally, hybrid manufacturers make use of lightweight materials, such as aluminum, so the vehicle needs less energy to accelerate.

Hybrid vehicles have other advantages. In 2003, 2004, and 2005, buyers of hybrid vehicles were entitled to a $2,000 federal income tax "clean fuel" deduction, the government's way of promoting interest in hybrid vehicles. As of 2005 that deduction was scheduled to be reduced and then phased out. Supporters of hybrids, naturally, were working to get legislation passed to extend the deduction.

As of 2005 at least fifteen states gave tax credits to hybrid vehicle buyers, and thirteen other states were considering doing so. Oregon offered a state tax credit of up to $1,500; Connecticut waived the 6 percent sales tax on the car, and Colorado offered a tax credit of up to $4,713. Hybrids can also go on some toll roads free. In some cities, such as San Jose and Los Angeles, California, hybrid car owners do not have to feed parking meters in city lots or on the streets. Some states allow hybrid cars with just the driver and no passengers to use car-pooling lanes. And some states release hybrids from emissions inspections. In London, England, hybrid vehicle owners pay the lowest amount of tax on their cars and do not have to pay a "congestion charge," a tax levied on all other vehicles in the city.

Types of hybrid vehicles

Hybrid vehicles come in two basic types: series and parallel. In a series hybrid, the first generation of modern hybrids, the gasoline-powered engine never powers the car directly. Rather, the gasoline engine turns a generator, which powers an electric motor that in turn powers the drive train, or it recharges the batteries. In a parallel hybrid, the second generation of modern hybrids, the gasoline-powered engine and the batteries power the car at the same time. In these cars, both the gasoline-powered engine and the electric motor are attached, independently, to the car's drive train. A third generation of hybrids is being developed. These vehicles use a differential-type linkage and a computer to allow the vehicle to be powered by the internal combustion engine, the electric motor, or both. The computer shuts off the gas engine when the electric motor is providing enough energy.

Other terms are frequently used to describe various sorts of hybrids. Sometimes the terms *strong hybrid* or *full hybrid* refer to the third-generation vehicles that can be powered by the gas engine, the electric motor, or both. The term *assist hybrid* refers to vehicles in which the battery and electric motor are used to accelerate the vehicle in combination with the gas engine. *Plug-in hybrids*, sometimes called *gas-optional* or *griddable*, have larger batteries and are able to run entirely on electricity from the electric motor and batteries. These vehicles can be recharged by plugging them into the power grid. The vehicle can rely on this electricity for short hops and daily commuting, but it also has a gas-powered engine for use during longer trips. *Mild hybrids* are often sold as hybrids, but they are not true hybrids because the electric motor never powers the vehicle. They are able to achieve greater fuel efficiency, however, because a starter motor spins the engine to the number of revolutions per minute it needs to operate before fuel is injected. These vehicles also use "regenerative" braking, and their engines do not run when the vehicle is coasting, braking, or idling.

The future of hybrid vehicles

As of 2005 only about one percent of new cars purchased were hybrids. In 2004, however, the number of hybrid registrations was up 81 percent from the year before, to just over 83,000. Many car industry observers believe that momentum is building in the hybrid industry and that consumer demand is growing enough to encourage manufacturers to design and build them. In the early 2000s the three hybrid cars available in the United States were the Honda Civic Hybrid, the Honda Insight, and the Toyota Prius. In designing its cars Honda aimed for the highest gas mileage possible, and its cars can get up to 68 miles (109 kilometers) per gallon. Toyota, on the other hand, aimed primarily for pollution reduction. The gas mileage of the Prius is in the mid- to high 40s.

Other car manufacturers made plans to release hybrid models. Scheduled for release in 2005 were hybrid vehicles from Daimler-Chrysler (Dodge and Mercedes), Ford, and General Motors (Chevy, GMC-Sierra, and Saturn). In the early 2000s many conservationists were growing concerned about the large and growing number of SUVs, which are classified as trucks and therefore are not required to get gas mileage as high as that of cars. Accordingly, some manufacturers are designing hybrid trucks and SUVs. In 2005 Toyota and Lexus were both planning to release hybrid SUVs, and Chevy scheduled offerings of two models of hybrid pickup trucks. Industry observers believe that the number of hybrids sold in 2005 could equal the total number sold in the four preceding years combined.

Some experts question the value of hybrid cars, at least from a strictly economic standpoint. While they support efforts to reduce pollution, they point out that, as of 2005, the higher price of hybrids offsets much of the energy savings. The magazine *Consumer Reports* calculated that it would take about 21 years of energy savings to offset the higher price of one popular hybrid model without the tax deduction. With the tax deduction, it would still take about four years for the buyer to break even. These estimates, however, assume that gas prices will remain consistent. In 2005, and early 2006, gas prices rose dramatically, thus making the payback period for hybrids shorter. For the near term, industry experts are also concerned about the resale value of hybrids, given that improvements are continually being made in the technology. Further, auto industry experts note that it is possible to achieve nearly similar energy savings with standard cars, some of which cost much less. Driving a stick shift vehicle as opposed to one with an automatic transmission can achieve gas savings of up to 18 percent.

Tips for more fuel-efficient driving

Though hybrid vehicles offer promise for reducing the U.S. reliance on fossils fuels, they are not the only solution. There are many other ways in which people can immediately reduce the amount of fossil fuels used by making personal choices to limit their own use of traditional automobiles. Drivers can also take a number of steps to increase the fuel efficiency of their existing vehicles or to use less fuel, whether the vehicle is a hybrid or not:

1. Use your legs. People can bicycle or even walk to many of their destinations, a solution that is better for both for the environment and an individual's health in general.

2. Utilize public transportation when possible. Public transportation is an option in many larger cities, though the structure of many U.S. cities (or their urban sprawl) needs to be addressed in others. One city bus can keep 40 or so vehicles off the road and save over 21,000 gallons (79,493 liters) of gasoline each year.

3. Car-pool. Car-pooling not only saves fuel by taking vehicles off the road, but it also reduces traffic congestion. Many cities encourage car-pooling with special lanes set aside for cars with two or more passengers.

4. Plan efficient trips. For long-distance trips, drivers can save fuel by taking the most efficient route, which may not necessarily be the shortest route. Taking a bypass around a city might add miles, but it eliminates the stop-and-go driving of cities and suburbs that uses more gasoline.

5. Avoid short trips when possible. A vehicle reaches its peak operating efficiency only after it has warmed up for a few miles. Short hops of under a few miles use more fuel per mile than longer trips. Drivers can save fuel by combining errands in the same trip. In winter, combining errands can also reduce the number of cold starts the car has to make.

6. Reduce quick accelerations and stop-and-go driving. Cars consume the most fuel when they are accelerating. Fast accelerations waste fuel, and racing up to stoplights or stop signs, applying the brakes, then racing on to the next stop is especially wasteful. By anticipating stops, coasting, then gently accelerating, drivers can save fuel. One test showed

that "jackrabbit" driving, or driving with quick starts and hard braking, saves only 4 percent of a driver's time (two-and-a-half minutes for a one-hour trip) but consumes 39 percent more fuel.

7. Slow down. Driving at 55 mph (89 kph) can produce gas mileage gains of 15 percent compared to driving 65 mph (105 kph).

8. Reduce idling. Most drivers tend to let their car idle when it is stopped. Many believe that it takes more gas to restart the car than is consumed by idling. Tests, however, show that this is not true if the idle time is more than about 10 seconds. Turning the car off when a long delay is anticipated (for instance, at a railroad crossing, when waiting to pick someone up, or when waiting in line at drive-through windows) can save significant amounts of fuel. Idling a car for long periods of time in order to warm it up in cold weather wastes fuel. The best way to warm a car up is by driving it—again, avoiding quick acceleration, especially when the car is cold.

9. Use engine-block heaters in cold weather climates. For a short trip a car can use up to 50 percent more fuel in cold weather than in warm weather. Plug-in engine-block heaters allow a car's engine to reach peak operating efficiency in cold weather much faster, saving fuel.

10. Reduce weight. Two ways to reduce weight are to clear snow and ice off the car, including ice that builds up in the wheel wells, and not to carry around excess items in the trunk or in the bed of a truck. Removing items such as ski racks and bicycle racks when they are not being used can increase fuel efficiency by reducing both weight and aerodynamic drag. Airlines discovered that they could save thousands of dollars per year in fuel costs solely by switching from glass to lighter plastic bottles for beverages on jumbo jets.

11. Keep tires inflated. A tire that is underinflated by just two pounds per square inch can increase fuel consumption by one percent. Tire pressure drops in cold weather, so it is especially important to check the tires' pressure in winter.

12. Service the vehicle. Such things as fouled spark plugs and dirty air filters can reduce fuel efficiency. A periodic wheel alignment can increase gas mileage by up to ten percent.

13. Use tires appropriate for conditions. Most city and suburban drivers do not need snow tires, which are softer and increase fuel consumption. For these drivers all-season radial tires are sufficient. On the other hand, drivers in rural areas, where roads can often be snow-packed, might achieve greater fuel efficiency with snow tires because they can reduce slippage.

14. Shift up. With manual transmissions shifting into a higher gear as soon as possible saves fuel. Manual-transmission vehicles get up to 18 percent better gas mileage than automatic-transmission vehicles.

15. Turn off the air-conditioner. Minimizing the use of air-conditioning, which is powered by the vehicle's engine, increases fuel efficiency, especially at lower speeds, when the amount of aerodynamic drag is not significantly increased by opening the vehicle's windows.

LEAVING AN ENERGY FOOTPRINT ON THE EARTH

Though innovation and creativity in creating energy efficient buildings, cars, and appliances and using renewable energy sources will begin to reduce the use of fossil fuels in the future, the choices that people make today, in their everyday lives, can also make a major difference. Every person on the planet leaves a "footprint" on the Earth, a demand on nature that includes the energy taken to support a person's consumption habits, whether they are choosing food, housing, utilities, transportation, or other goods and services (like clothing, recreation, and cleaning products). For the Earth to support a growing population, there must be a balance between increasing human needs and wants and nature's ability to sustain all of the energy requirements placed on the planet's resources.

There are daily actions beyond those mentioned above that everyone can take to reduce the overall energy footprint on the Earth and to help stay within the planet's capacity for regenerating energy, food, and materials, including:

1. Limiting excess consumption. People can make choices not to buy items they do not need, to purchase recycled or secondhand items, or, if a new product is necessary, to purchase non-disposable items that require little or no packaging. Energy is used to make any product; thus, the fewer products people buy, the less energy is used. Many times, a secondhand item, especially a durable good such as a piece of furniture, is just as good as purchasing a new

item, and will be cheaper for the consumer as well. The excess packaging of items affects energy consumption in several ways, including in the production of the packaging itself and the fuel necessary to power the dump trucks that must take the additional garbage to the landfill. Rather than buying a product that is contained in both a box and a plastic bag, people can choose items that have less packaging or are not packaged at all.

2. Recycling. Many items that go into landfills can be recycled into products to save energy. Waste paper can be made into insulation. Plastic can be turned into a host of products, such as durable carpeting for use in cars. In the United States, about 250 million automotive tires are scrapped each year. In 1989 only about 10 percent of those tires were recycled, but by the 2000s, that percentage had increased to 80 percent. The tires are commonly shredded to provide fill in building

A technician holding a handful of shredded waste plastic at a recycling facility. These shredded fragments will be heated to over 750°F (400°C) in a fluidized bed of sand. This breaks down the plastic into its basic hydrocarbon constituents, which are given off as gases. These are filtered and cooled to produce a very pure, waxy substance which can be used by oil refineries and the petrochemical industry. *J. King-Holmes/Photo Researchers, Inc.*

Bales of HDPE (high density polyethylene) plastic is inspected at a recycling plant. HDPE is used in a variety of rigid packaging for food and beverages. It is sorted and compressed into these bales before being cleaned and shredded. It is then shipped in chipped form to manufacturing plants for re-use. *Hank Morgan/Photo Researchers, Inc.*

projects, mulch for gardens and under playground equipment, and as an ingredient in road asphalt. Soda pop bottles can be recycled to make the synthetic fleece common in winter jackets. Aluminum cans are 100 percent recyclable. In 2003, 54 billion aluminum cans were recycled, saving an amount of energy in aluminum manufacture equal to about 15 million barrels of crude oil.

3. Lighting. Artificial lighting is a major consumer of electricity. Energy can be saved by turning off lights that are not in use, relying on natural lighting during the day, and attaching motion sensors to outdoor lighting so that the lights come on only when they are needed for outdoor activity. The use of compact fluorescent lightbulbs throughout a home can cut energy usage for lighting by about three-fourths.

4. Food choices. Eating a diet with fewer animal-based and more plant-based products generally requires less energy, land, and other resources. Planting a garden or choosing

locally grown goods rather than buying items that must be transported cuts down on energy and pollution from shipping, packaging, fertilizers, and pesticides. Buying items that are not processed saves energy used in the canning, freezing, or packing industries.

Even if it seems that the actions you take are small and will not affect the planet, your contributions over your lifetime can make a difference. You can also ask your parents, other family members, and friends to take some of the steps listed above, and work together to encourage changes in energy efficiency and conservation.

■ ■ ■

For More Information

Books

Frej, Anne B. *Green Office Buildings: A Practical Guide to Development.* Washington, DC: Urban Land Institute, 2005.

Husain, Iqbal. *Electric and Hybrid Vehicles: Design Fundamentals.* Boca Raton, FL: CRC Press, 2003.

Hyde, Richard. *Climate Responsive Design.* London: Taylor and Francis, 2000.

Kibert, Charles J. *Sustainable Construction: Green Building Design and Delivery.* New York: Wiley, 2005.

Wulfinghoff, Donald R. *Energy Efficiency Manual: For Everyone Who Uses Energy, Pays for Utilities, Designs and Builds, Is Interested in Energy Conservation and the Environment.* Wheaton, MD: Energy Institute Press, 2000.

Periodicals

Feldman, William. "Lighting the Way: To Increased Energy Efficiency." *Journal of Property Management* (May 1, 2001): 70.

Motavalli, Jim. "Watt's the Story? Energy-Efficient Lighting Comes of Age." *E* (September 1, 2003): 54.

Web sites

"Driving and Maintaining Your Vehicle." Natural Resources Canada. http://oee.nrcan.gc.ca/transportation/personal/driving/autosmart-maintenance.cfm?attr=11 (accessed on September 28, 2005).

"Ecological Footprint Quiz.;rdquo; Earth Day Network. http://www.earthday.net/footprint/index.asp (accessed on February 6, 2006).

"Energy Efficiency and Renewable Energy." U.S. Department of Energy. http://www.eere.energy.gov (accessed on September 28, 2005).

"Green Building Basics." California Home. http://www.ciwmb.ca.gov/GreenBuilding/Basics.htm (accessed on September 28, 2005).

"Incandescent, Fluorescent, Halogen, and Compact Fluorescent." California Energy Commission. http://www.consumerenergycenter.org/homeandwork/homes/inside/lighting/bulbs.html (accessed on September 28, 2005).

"Introduction to Green Building." Green Roundtable. http://www.greenroundtable.org/pdfs/Intro-To-Green-Building.pdf (accessed on September 28, 2005).

Nice, Karim. "How Hybrid Cars Work." Howstuffworks.com. http://auto.howstuffworks.com/hybrid-car.htm (accessed on September 28, 2005).

"Thermal Mass and R-value: Making Sense of a Confusing Issue." BuildingGreen.com. http://buildggreen.com/auth/article.cfm?fileName=070401a.xml (accessed on September 28, 2005).

Venetoulis, Jason, Dahlia Chazan, and Christopher Gaudet. "Ecological Footprint of Nations: 2004." Redefining Progress. http://www.rprogress.org/newpubs/2004/footprintnations2004.pdf (accessed on February 8, 2006).

Possible Future Energy Sources

The word "energy" fills the pages of this book, and many forms of energy are described in previous chapters. Energy is, however, really a very tricky and difficult term to define exactly, even for students who have studied physics and engineering for many years. While it is very useful to think of energy as something that can flow like a river from one thing to another, or be stored in a battery, energy really isn't a "thing" at all. Energy does not exist by the gallon or liter, but a gallon or liter of gasoline has a certain amount of energy, an ability to flow through the process of combustion inside a properly designed engine, to turn gears and wheels that move a car.

Scientists and engineers usually describe or define energy as an object's ability to do work, to move things, make things hotter, and so forth. For example, the sun does not transfer a substance called energy to Earth. The nuclear reactions in the sun produce light that travels through space and that increase the energy level of objects the light strikes on Earth. For example, the light strikes objects and the light's own energy or ability to do work then changes molecules in the object that allow them to undergo chemical reactions or make them move and thereby cause the object's temperature to rise. As students advance in their studies their understanding of energy will change.

When thinking about the possible sources of energy to be used in the future, however, it is important to keep in mind that because energy is not a thing itself, but rather something that everything *has*, we can look for potential sources of useable energy. The world

Words to Know

Cold fusion Nuclear fusion that occurs without high heat; also referred to as low energy nuclear reactions.

Electromagnetism Magnetism developed by a current of electricity.

Fusion The process by which the nuclei of light atoms join, releasing energy.

Heisenberg uncertainty principle The principle that it is impossible to know simultaneously both the location and momentum of a subatomic particle.

Magnetic levitation The process of using the attractive and repulsive forces of magnetism to move objects such as trains.

Perpetual motion The power of a machine to run indefinitely without any energy input.

Superconductivity The disappearance of electrical resistance in a substance such as some metals at very low temperatures.

Thermodynamics The branch of physics that deals with the mechanical actions or relations of heat.

Tokamak An acronym for the Russian-built toroidal magnetic chamber, a device for containing a fusion reaction.

Zero point energy The energy contained in electromagnetic fluctuations that remains in a vacuum, even when the temperature has been reduced to very low levels.

will eventually run out of substances such as oil that can be found and used at a reasonable cost, but the Earth will never run out of energy. The challenge for future generations is the ability to harness and use new sources of energy to do work.

IS ALTERNATIVE ENERGY ENOUGH?

Overuse of fossil fuels such as coal, natural gas, and petroleum as a source of energy can cause pollution, mining damage, and contribute to climate change. They are increasingly limited resources that because they are valuable, can even become a cause of war. Regardless of attempts to make cars, machines, and devices that use fossil fuels more efficiently, fossil fuels will someday be very scarce and hard to find. The world needs other energy sources that are clean, renewable, and affordable.

Most sources of "alternative" energy—which usually means energy from any source other than fossil fuels and nuclear fission—depend on obvious, natural sources of energy. The sun bathes Earth with light, which can either be turned into electricity or used directly for light or heat. The wind and rivers are loaded with kinetic energy (the energy of matter in motion). Tides raise and lower the sea, and hold a potentially useable source of energy.

There is nothing new about these energy sources. People have always used the sun to light spaces, dry food and clothing, and heat buildings. Water wheels and windmills have done useful work for centuries. The challenge for modern scientists and engineers, however, is to find effective ways of harnessing these power sources (and others) on a scale large enough and a cost low enough to meet the needs of the more than six billion people already living on Earth, a number that is expected to increase.

Many alternative or renewable energy sources, especially hydroelectric power, wind, and solar power, are already providing important amounts of energy or are capable of providing significant amounts of energy in the near future. These energy sources have many advantages over fossil fuels, but they also have limitations. One problem with some of them is that to provide truly large amounts of energy, they require huge, expensive facilities. Hydroelectric power needs massive dams that drown land, displace towns and villages, and threaten wildlife habitats (the living environment). Tidal or wave power needs dams across tidal basins and machines for gathering wave energy, all of which would not only be expensive but might spoil wild shorelines and disturb sea life. Solar cells to turn sunlight into electricity are getting steadily less expensive, but a solar power plant big enough to make as much energy as a coal or nuclear plant would cover a large area of land. Today, large windmills can make electricity more cheaply than either coal-burning plants or nuclear power plants, yet wind farms consist of large numbers of towering windmills—often twice the height of the Statue of Liberty—that change landscapes and can kill birds with their whirling blades. In addition, people often need more electricity than can be produced or stored while the sun is shining or when the wind is blowing.

Nuclear power plants produce steady-flowing energy, but not all experts agree that building many new nuclear power plants would be an affordable way to meet the world's energy needs. Quite apart from possible problems like radioactive waste, potential terrorist attacks on reactors, or reactor accidents, nuclear power has always been—and, according to some experts, still is—more expensive than other energy sources. Contrary to popular belief, for example, orders for nuclear plants practically stopped in the United States *before* the near-disaster at the Three Mile Island nuclear power plant in Pennsylvania in 1979. Nuclear plants were simply too expensive. And they have remained so. Since 1973, orders for new nuclear power plants in the United States have consistently been cancelled. The last non-military nuclear reactor to start operations in the United States was at the Watts Bar nuclear power plant in Tennessee in the late 1990s.

But nuclear power is not the only energy source with problems. Large, centralized renewable-energy projects must be placed in specific geographic locations and may damage the environment. A hydroelectric dam needs to be built on a river, for example, and many rivers have already been dammed in some way. A wave-power or tide-power generating station would have to be built on a specific type of ocean shoreline. Windmills need strong, reliable winds, which are not found everywhere. Solar power does best with steady sunshine, as in deserts and the tropics. Only in certain places is geothermal heat is close enough to the Earth's surface to be useful. There is really not one electrical energy problem but two: the problem of generating electricity and the problem of transporting electrical power.

So, while the energy of the wind, sun, oceans, and atoms is inexhaustible, our ability to capture it is limited by geography, money, safety, and other considerations. In fact, experts argue that these sources of power will never be able to safely, cleanly, and affordably supply the world with all the energy it needs. Further-more, all the sources of energy mentioned so far in this chapter are sources of *electricity*, but not all our energy needs can be met by electricity. Heating buildings with electricity is very expensive, and electric cars and trucks that can compete with the power and speed of fossil-fuel-powered vehicles do not yet exist. Electricity, whether it comes from windmills or nuclear power, cannot help us to break our addiction to the liquid fossil fuel known as "oil,"—petroleum, from which gasoline and other fuels are made.

However, defenders of new energy sources have at least possible answers to many problems and objections. Just as advocates of nuclear power argue that with new reactor designs, nuclear power can be made safer and cheaper, supporters of windmills and solar power argue that new designs will eliminate limitations of these technologies. For example, large windmills might coexist with ranching on the wide-open landscapes of the American Midwest or be located far out to sea, while smaller, more efficient, vertical-axis windmills (which resemble upside-down eggbeaters and do not harm as many birds as other designs) can be placed on roof-tops. Solar panels can also be placed on rooftops, producing power where it is needed without using more land. And by using elec-tricity from windmills or solar panels to break water (H_2O) into

hydrogen and oxygen and then using the hydrogen in fuel cells (a type of chemical battery) to make electricity, we can get power from the wind and sun even when the wind is not blowing or the sun is not shining. Hydrogen can also power cars and trucks, and biofuels may also help fuel vehicles.

As for whether renewable energy sources can make all the energy that modern civilization needs, many experts argue that by using energy more efficiently we can reduce our energy demand to the point where we can rely on what renewables can give us without giving up any of the advantages of a high-technology lifestyle. Some experts also argue that nuclear power will, in fact, be necessary. This remains a controversial subject.

But apart from increasing efficiency—which has already reduced energy use for many tasks and could reduce energy usage much more—no alternative perfect solutions are yet available. Solar panels are still too expensive to put on every rooftop (though

The world's most technically advanced solar power facility, Solar Two, focuses the sun's rays on a tower containing liquid salt, which is pumped into an insulated tank so that stored heat can drive turbines and provide electric power 24 hours a day.
© George Steinmetz/Corbis.

Japan and Germany, with their huge government-backed solar programs, may be changing that). Claims of greater safety and lower cost for new nuclear power-plant designs are still just promises. Vertical-axis windmills have not yet been widely installed or tested. As of early 2006, the closest thing to an alternative-energy "revolution" is what is happening in wind power: large windmills have been the cheapest, mostly rapidly-growing source of new electricity worldwide since the early 2000s. Yet some people are objecting to plans to build large wind farms in visible or fragile locations, such as the mountaintops of Vermont or off the coast of Massachusetts. Windmills are still making only a small fraction of our electricity, and until it is affordable to use them to make hydrogen for fuel cells on a large scale—which it is not, yet—we will not be able to obtain most of our electricity from wind no matter how many windmills we build.

DREAMS OF FREE ENERGY

Many of these problems with alternative energy sources will undoubtedly eventually be solved. In the long run, some mixture of wisely-used alternative sources could power our civilization for as long as need be. Yet some people still dream of very inexpensive, inexhaustible energy from exotic or unproved sources. Nuclear power itself began as one such dream. In 1954, the chairman of the U.S. Atomic Energy Commission said in a speech that that "it is not too much to expect that our children will enjoy electrical energy too cheap to meter." (Metering is the process of measuring how much a given amount of electricity costs.) Some scientists even predicted that small nuclear plants would someday power individual homes, cars, and airplanes. Those dreams or predictions did not come true, mostly because nuclear energy still requires dangerous and complex technology. Far from being too cheap to meter, nuclear power is as expensive as any other standard way of making electricity.

But could some other technology, something completely new, fulfill the dream of cheap, endless power? Most energy experts and engineers urge us not to expect an energy miracle, and to be prudent in the use of resources we have and know, but scientists and inventors continue the search for new sources of energy.

Some of the methods that have been proposed for making cheap, endless power have no scientific basis and are simply "fake science" ("pseudoscience.") The most famous of these fake energy sources is perpetual motion. Some other proposed methods, such

Johann Bessler and the Bessler Wheel

One of the most famous figures in the dubious history of perpetual motion was German engineer and inventor Johann Bessler (1680–1745). In 1712 Bessler unveiled his first machine, called the Bessler wheel, which he claimed was a perpetual motion machine that drew its power from gravity. Throughout his career, Bessler attempted to sell the machine, wanting the money to establish a Christian-based school of engineering. He never found any buyers and he refused to reveal the "secret" of the machine until he was paid. Never able to find a buyer for his machine, he died in poverty without having revealed the "secret" of the Bessler wheel.

Skepticism (a preexisting doubt in the truth of a matter) was increased when one of Bessler's maids testified that she and other servants were manually turning the wheel with a crank from another room, which was attached to the wheel by a rod and series of gears.

Bessler allegedly encoded the "secret" of his perpetual motion machine in the text of his books, including *Apologia Poetica* (*Poetic Defense*, published in 1716), *Das Triumphirende Perpetuum Mobile Orffyreanum* (*The Triumphant Orffyrean Perpetual Motion*, published in 1719), and *Maschinen Tractate* (*Tract on Machines*, published in 1722).

as "zero point energy," have some slight scientific basis, but most scientists still think they are not useable given human and Earth's own limitations. Still other possible energy sources (such as cold fusion and sonofusion) are studied seriously by a number of real scientists, but the majority are still not convinced that they can produce useable energy in the foreseeable future. Finally, there are some methods that all scientists agree are physically possible, such as hot fusion and solar power satellites, but many experts do not agree that these schemes will ever be practical. A number of possible cheap-energy schemes, from the silly to the serious, are discussed in the rest of this chapter.

PERPETUAL MOTION, AN ENERGY FRAUD AND SCAM

Some people argue that a "perpetual motion machine" can be built that will produce endless energy without having to burn fuel or harvest energy from an outside source such as the wind or sun. The search for such a magical machine dates back to at least the thirteenth century, when French artist Villard de Honnecourt drew fanciful pictures of perpetual motion machines. Since that time, many inventors and tinkerers have sought, without success, to design a machine that produces energy without any need for energy to be put in. Some fake perpetual motions have even been built to fool the public or steal money from investors.

Any so-called perpetual motion machine would violate laws of thermodynamics, which places limits on the nature and direction of heat transfer and the efficiencies that can be achieved by any type of system. This means it is impossible to construct any such system or machine with 100% efficiency. An important but complex part of the laws of thermodynamics is termed *entropy*. Entropy essentially means that without the use of energy, all systems or machines must move to disorder (experience decayed or diminishing efficiency) over time. Accordingly, the only way anything can be perpetual is to use energy to maintain the system or machine. Any statement to the contrary (against) violates the laws of physics.

Despite the claims of scam artists or "inventors," scientists agree that perpetual motion can never be an energy source. It is impossible to get more energy out of a machine than you put into it: machines can only change the form that energy is in. The laws of physics say that you can't get something for nothing—at least, not for long. In a sense, it is possible to store "energy" for a while in some devices, but batteries and other storage devices (which also decay over time) can only give back whatever energy is put into them. Perpetual motion machines will never supply the world with energy.

ADVANCES IN ELECTRICITY AND MAGNETISM

As scientists continue to explore the nature of electricity and magnetism (actually different aspects of a combined fundamental force appropriately termed electromagnetism) so, too, are engineers advancing ways to convert this knowledge into useable forms of energy, and to improve the efficiency of power transmission, transportation, and so forth. Although improved efficiency does not provide new energy, it can have the same impact as developing new sources because it allows existing sources to do more things or last longer.

Magnetism

People have known about the power of magnetism for thousands of years. In ancient Greece, near the city of Magnesia, mysterious stones with the power to attract iron were first discovered. Later, the Chinese discovered that if one of these stones was stroked with a needle, the needle became magnetic. Around the year 1000 the Chinese discovered that when such a needle was suspended, it would point in the direction of the North and South Poles. The result of this discovery was the magnetic compass, which helped to open the world's oceans to navigation and exploration.

It was not until the nineteenth century that physicists began to understand magnetism and magnetic fields. Essentially,

A reconstruction of an experiment that Hans Christian Oersted (1777-1851) constructed to show that elecromagnetism can be produced by an electrical current. © *DK Limited/Corbis*.

magnetism is a force that attracts such substances as iron, but also cobalt and nickel, at a distance. What causes the attraction is decribed by lines of flux ("lines" on a plane that cross or include magnetic poles) that come from electrically charged particles that spin. These lines flow from one end of an object to the other. The ends are commonly referred to as the north and south poles, similar to the terms applied to Earth's poles. In a magnetic field, the flux flows from the north to the south. While individual particles such as electrons can have magnetic fields, so can larger objects, such as the magnets that hold notes and shopping lists to the door of a refrigerator. When an object with a magnetic field exerts its force on another object with a magnetic field, the result is magnetism.

The north pole of one magnet attracts the south pole of another and, conversely, the north pole (or south pole) repels the north pole (or south pole) of another magnet. The lines of flux cause this attraction or repulsion. Just as these lines flow from the north to the south of one object, they can flow from the north of one object to the south of another, pulling the two objects together, almost like two spinning gears in a car that mesh smoothly together. When like poles—for example two north poles—are brought together, the lines of flux are flowing in opposite directions, causing the two objects to, in effect, bounce off each other, like two spinning tops that collide and bounce away.

Electromagnetism

In the twenty-first century magnetism powers devices such as tape drives, speakers, and read/write heads for computer hard drives. The energy is captured through electromagnetism, which is based on the simple principle that an electrical current, which consists of a flow of electrons passing through a wire, creates its own magnetic field. This magnetic field moves in a direction perpendicular to the flow of the current in the wire. This force is called the Lorenz force, named after Dutch scientist Henrick Antoon Lorenz (1853–1928).

A simple electromagnet can be created with a battery and a piece of wire. If the wire is connected to the positive and negative poles of the battery, the electrons collecting at the negative pole will "flow" through the wire to the positive pole, rapidly depleting, using up, the battery. Generally, something is attached to the middle of the wire—a radio, a lightbulb, a toaster—so that the electricity can do work while at the same time offering resistance

so the battery does not quickly go dead. The magnetic field of a single strand of wire, however, is likely to be relatively weak, because the Lorenz force weakens as the distance from the wire increases. One way to strengthen the magnetic field is to coil the wire, in effect recruiting multiple strands of wire to create a magnetic field that pulls (or pushes) in the same direction. The more coils of wire, the stronger the magnetic field.

This is the basic science behind magnetic levitation. In its application, magnetic levitation is a process by which train cars are "levitated," or raised, so that rather than riding on tracks, they ride on a cushion of air. The chief advantage of "maglev" trains is that this cushion of air, combined with the trains' aerodynamic design, virtually eliminates the energy lost because of friction. The result is lower cost per operating mile and lower maintenance costs because of less wear and tear on the equipment. Although exact estimates of savings vary, the operating cost of a maglev train in

The magnetic compass was one of the first uses of magnetic properties. Maglev trains are also based on the power of magnets.
© *Matthias Kulka/zefa/ Corbis.*

Shanghai's $1.2 billion maglev train arrives at Long Yang station after its 267 mph (430 kph) trip from Pudong Airport in Shanghai, China, June 7, 2005. © *Mark Ralston/Reuters/Corbis.*

terms of cost per passenger mile traveled is only a fraction of the cost of auto and air transport.

ZERO POINT ENERGY

Zero point energy sounds like magic or science fiction: energy that comes straight out the vacuum of empty space. A handful of scientists argue that zero point energy can be harnessed to provide power. Most scientists, however, are very skeptical (doubtful) that it can ever be turned into a practical power source.

The idea of the nature and potential of a vacuum has long interested scientists. In ancient Greece, the philosopher Aristotle (384–322 BC) argued that "nature abhors a vacuum." That is, he taught that it was impossible for any region of space to be totally empty. For almost two thousand years, scientists accepted Aristotle's teachings, but by the middle of the seventeenth century they had come to reject it. In 1644, an Italian scientist named Evangelista Torricelli (1608–1647) invented an early barometer, a standing glass tube filled with mercury. The top of the tube was sealed and the bottom curved back up to an opening so

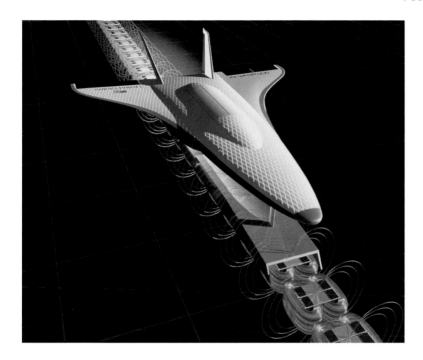

Computer-aided design (CAD) of a spacecraft being launched by maglev. This system uses magnets to float a vehicle along a track, and will reduce the cost of space travel. It could help launch a spacecraft from an airport runway to orbit every 90 minutes. *NASA/Photo Researchers, Inc.*

that the atmosphere could push on the exposed mercury. When the pressure of the atmosphere rose or fell due to the weather, it would push on the mercury with changing pressure, causing it to rise and fall in the glass tube. Torricelli noticed that even if the tube was made without air above the mercury, an open space would appear there. Because air cannot pass through mercury, Torricelli reasoned that this empty space at the top of the tube had to be a true vacuum—a volume of space containing no matter. In later experiments, other scientists confirmed his arguments.

For several hundred years after Torricelli, scientists argued that a vacuum was a region of space in which "nothing" existed. In the early twentieth century, however, physicists discovered the strange properties of matter that are obvious only for very small objects such as atoms and electrons. The new knowledge, called quantum physics, forced scientists to question whether the vacuum was in truth entirely empty. It became clear that Aristotle had been right (though for the wrong reasons), and that there is really no such thing as empty space. (In physics, "space" does not mean outer space, but rather any volume, including the space inside an atom, a bottle, or a room.)

A woodcut, ca. 1850, of Otto Von Guericke's experiment with the Magdeburg Hemisphere demonstrating the pressure of air. © *Bettmann/Corbis.*

Quantum physics is that branch or subdivision of the study of physics that started with the observation that an atom is like—and yet unlike—a tiny solar system. The atom's nucleus—a very small object or particle, much heavier than anything else in the atom—is positioned like a microscopic "sun,", and electrons, many times smaller than the nucleus, orbit it in some ways like tiny "planets." A question that puzzled physicists in the nineteenth century was why the orbiting electrons of an atom do not quickly radiate away their energy in the form of light and fall into the nucleus. On the contrary, they *never* do so. To explain this fact, modern physicists developed quantum physics, which explains matter and energy as having both wave- and particle-like features. They found that energy does not flow smoothly, but always changes in small jumps or fixed quantities. They called each of these jumps a "quantum", thus giving the new physics its name, "quantum physics." Quantum physicists showed that electrons orbiting an atom's nucleus are not really like tiny planets at all, except as we may picture them in our minds.

The German physicist Werner Heisenberg (1901–1972) deepened our understanding of quantum physics in 1927, when he announced what is now called the "uncertainty principle." The uncertainty principle states that by the very nature of matter and energy, it is impossible to measure everything about an object with perfect accuracy. For example, the better one's measurement of the position of an electron gets, the poorer one's knowledge of its momentum (a measure of both mass and velocity) gets, and that the reverse is also true because the better the understanding of momentum, the less one can know about position. There is no way to make better measurements: as Heisenberg proved, the uncertainty or lack of ability to know is not a form of ignorance, but arises from the nature and laws of the universe itself. It isn't that we don't know what the precise values are; the precise values simply don't exist.

According to the uncertainty principle, which has been tested many thousands of times in laboratories, there is a certain amount of fuzziness or uncertainty about all physical phenomena. This includes the vacuum. In fact, the uncertainty principle says that there can be no such thing as a perfect vacuum. Perfect emptiness or vacuum would mean that there was zero matter and energy, but "zero" is a precise value, and absolutely precise values are forbidden by the nature of the universe.

Instead, physicists now know that "virtual" particles are continuously popping into and out of existence everywhere, throughout all space, including the "vacuum"—the apparently empty space found between atoms and stars, and also at the top of Torricelli's glass tube. These virtual particles include photons (particles of light). All particles and waves are forms of energy, as German scientist Albert Einstein (1879–1955) proved in 1905, so the existence of virtual particles means that the "vacuum" is boiling invisibly with energy all the time, everywhere. This energy is called "zero point energy." A few physicists—but not most—argue that zero point energy can provide energy for human use.

A physicist working in the field of zero point energy, Dr. Hal E. Puthoff of the Institute for Advanced Studies in Austin, Texas, explains zero point energy in these terms:

> When you get down to the tiniest quantum levels, everything's always 'jiggly.' Nothing is completely still, even at absolute zero. That's why it's called 'zero point energy,' because if you were to cool the universe down to absolute zero—where all thermal motions were frozen out—you'd still have residual [leftover] motion. The energy associated with that 'jiggling' will remain, too.

Absolute zero temperature then is not zero energy, but the minimum energy that can exist.

A Childhood Genius

When Werner Heisenberg was in his early teens, another, older student needed a calculus tutor. Heisenberg had not studied calculus, because it was not taught at his school. So he taught himself calculus so that he could tutor the older student—while also practicing to become an accomplished musician. Heisenberg won the Nobel Prize for physics in 1932 and, besides his scientific work in quantum physics, wrote many books about the relationship of physics to philosophy.

Scientists agree that zero point energy is real. This energy cannot usually be felt or easily measured because it surrounds everything equally. Thus, its forces in effect cancel one another out, exerting pressure in all directions at once, just as the pressure of the Earth's atmosphere can't be felt because it pushing on the outside of your chest and on the inside of your lungs at the same time.

Puthoff is one of the scientists who argue that the amount of zero point energy in the vacuum is very large. "It's ridiculous," he says, "but theoretically, there's enough [zero point] energy in the volume of a coffee cup to more than evaporate all the world's oceans. But that's if you could get at all of it, and you obviously can't."

Whether the zero point energy is useable is a question, but it is certainly there. Physicists have measured a number of effects that prove its existence. One is called the Lamb effect or Lamb shift, named after physicist Willis Lamb (1913–). The Lamb effect refers to small changes in light given off by an excited atom. This is predicted as a side effect of zero point energy.

A more impressive demonstration of zero point energy is the Casimir effect, measured in 1948 by Dutch physicist H. B. G. Casimir (1909–2000). Casimir showed that if two metal plates are brought very close together, they attract each other very slightly. As the plates are drawn or pushed together (whether to describe it as drawn or pushed depends on the exact explanation of zero point energy used), it is at least potentially possible to extract energy from their motion.

So not only is zero point energy real, physicists agree that it can be made to do work. But they do not agree that zero point energy can ever be made to do enough work to be useful. The fact that something happens in the realm of quantum physics doesn't prove that it can be made to happen in the world of everyday objects.

A few scientists explore the idea that zero point power sources might make interplanetary space travel practical, for a spacecraft would be able to extract the energy it needs from the vacuum of space rather than having to carry fuel. Some science fiction writers (and some scientists too) envision a day when zero point energy could power fighter planes flying at four times the speed of sound, power 1,200-seat airliners flying at altitudes of 100 miles (161 kilometers) and covering 12,000 miles (19,312 kilometers) in 70 minutes, and power spacecraft making 12-hour trips to the moon.

But most scientists do not accept that such science-fiction scenarios are possible. It would take billions of Casimir plates to produce a useful amount of power, and more energy would be consumed in constructing and positioning of the plates than the plates could ever produce. Such a machine would use more energy than it made. Therefore, many scientists debate whether money and time spent on zero point energy research should be spent instead on research of other forms of energy. Nevertheless, it is true that unlike mechanical perpetual motion machines, zero point energy is studied by some real scientists. Physicists agree that *some* energy can be had "for nothing" from the vacuum. It is simply a question of how much.

FUSION

Fusion powers the sun and all other stars. Fusion is, however, very different from the process of generating nuclear power that is used in today's nuclear power plants. These are powered by nuclear fission, meaning that they release energy by splitting atoms apart into smaller atoms ("fissioning" them). This energy is used to turn water into steam, and the steam is used to turn generators that make electricity. Fusion, on the other hand, produces heat by "fusing" atoms, forcing them to come together into larger atoms. Fusion power, unlike fission, would produce only small amounts of radioactive waste and its fuel would not be dangerous to people's health.

But fusion does not yet produce useful power on Earth. For fusion power to be practical, scientists have to figure out how to

Lasers are focused on a small pellet of fuel in an attempt to create a nuclear fusion reaction for the purpose of producing energy. The fuel pellet undergoes a fusion reaction while being bombarded by the light from 24 lasers. © *Roger Ressmeyer/Corbis.*

make fusion happen in a steady, small-scale way, producing neither a fizzle nor an explosion. However, this has turned out to be a difficult trick. Billions of dollars have been spent over the last forty years trying to make fusion work, and success is still decades away—or may never be achieved. Three kinds of fusion research are described below: conventional or "hot" fusion, "cold" fusion, and sonofusion.

Conventional fusion

In nuclear fusion, the nuclei of two light atoms (such as helium or hydrogen, the lightest atoms) bind together to form a single heavier nucleus. For example, the nuclei of two ordinary hydrogen atoms, each of which is simply a proton (a positively-charged particle), merge to form the nucleus of a deuterium atom, which is a neutron and a proton bound together. (A neutron is a particle that weighs about the same as a proton but has no electrical charge. Deuterium is also a kind of hydrogen.) When a deuterium nucleus or other particle is formed by the coming together of smaller particles, its mass is generally less than the total mass of the original particles before they came together. The mass that seems to have disappeared has been released in the form of energy. The amount of this energy can be calculated by using Albert Einstein's famous equation, $E = mc^2$, which means that when the correct units are used, energy (E) is equal to mass (m) times the speed of light (c) squared. Not much mass has to "disappear" for the amount of energy released to be very large. This is because the speed of light is so large: about 300,000 kilometers per second (186,000 miles per hour).

Fusion reactions occur naturally throughout the universe. For example, scientists have learned that the primary component of stars is hydrogen gas. Over time, this hydrogen is turned by fusion into the gas named helium, as the nuclei of four hydrogen atoms combine to form one helium nucleus. Many other fusion reactions take place in stars. In fact, all the heavier elements of which Earth (and our own bodies) are made, such as carbon, iron, oxygen, silicon, aluminum, and uranium, are produced by the fusion of lighter elements in stars.

For fusion to happen, "electrostatic repulsion" must be overcome. Particles with the same electrical charge repel or push each other apart. Electrons have negative charge, protons have positive charge. The closer two negative charges or two positive charges get to each other, the harder they repel and the harder it gets to bring them any closer. If two protons are to fuse together to form a single nucleus, therefore, they must be thrown together at high speed. Where does that energy come from?

It comes from heat. Heat is merely the motion of atoms and molecules. The hotter a piece of metal is, for example, the faster the atoms in it are vibrating. The atoms in a hot gas shoot around freely like balls on a pool table, only much faster. The hotter a gas gets, the faster its particles go. As a gas is heated, for instance, its atoms move with faster and faster, so they collide harder and harder. When the collisions are hard enough, the nuclei of the colliding atoms may fuse, or join together. This type of reaction is called a "thermonuclear" reaction, from the Greek *thermo*, meaning heat.

The temperature needed for this type of fusion to take place is extreme, on the order of tens or hundreds of millions of degrees. This kind of heat can be found in the centers of stars, including the sun, but does not occur naturally on Earth. It does occur artificially on Earth, however, in fusion laboratories and hydrogen bombs.

Mike

The detonation of the first thermonuclear bomb, codenamed "Mike," took place on November 1, 1952, on the Eniwetok atoll, a small coral island in the Pacific Ocean. The U.S.-built bomb consisted of a cylinder 20 feet (6 meters) tall and 6 feet, 8 inches (2 m) in diameter, weighing 164,000 pounds (61,212 kg).

Even the bomb's designers were amazed by its explosive force. Its fireball was 3 miles (4.8 km) wide. Within ninety seconds, the mushroom cloud had risen 57,000 feet (17 m) into the air. Eventually, after five minutes, the cloud reached a height of 135,000 feet (41 m), with a "stem" eight miles (13 km) across. People on ships 100 miles (161 km) away saw the flash. The explosion completely destroyed the island of Elugelab, carving out an underwater crater that was 6,240 feet (1,902 m) wide and 164 feet (50 m) deep and lifting 80 million tons of soil into the air. A bomb of this type would devastate any city on Earth.

Fusion in bombs

Just as the fission process used in nuclear power plants was first applied to make bombs, like the fission bombs used by the United States to bomb the Japanese cities of Hiroshima and Nagasaki in 1945 to end World War II, so was fusion.

To create a hydrogen bomb, scientists begin with a quantity of hydrogen. To create a fusion explosion, the hydrogen must be heated until it is as hot as the core of a star. This is done using a fission bomb. The basic design for a hydrogen bomb, then, is to pack hydrogen in a container around a fission bomb. When the fission bomb explodes, it heats the hydrogen enough to start runaway fusion explosions. This fusion explosion can be tens or even thousands of times more powerful than the fission explosion that started it.

The fission bomb that was dropped on Hiroshima was a 20-kiloton bomb, meaning that it had an explosive force equal to that of 20,000 tons of TNT (a chemical explosive). The first fusion bomb exploded with a force equal to that of 10.4 million tons of TNT—some 500 times the power of the Hiroshima bomb. The largest hydrogen bomb ever exploded had a force equal to 50

The mushroom cloud from Mike, one of the largest nuclear blasts ever, during Operation IVY. The blast completely destroyed Elugelab Island. © *Corbis.*

million tons of TNT, about 2,400 times the explosive power of the bomb dropped on Hiroshima.

Controlled fusion

Just as they did after the first fission bombs were developed in World War II, scientists began to seek ways to provide peaceful energy with nuclear fusion. The basic process they focused on made use of two forms (isotopes) of hydrogen. (An isotope is a form of an element having fewer or more neutrons in its nucleus than other forms of the same element.) These isotopes, known as "heavy hydrogen" because they contain extra neutrons, are called deuterium and tritium. A normal hydrogen atom's nucleus consists of a single proton, but the nucleus of a deuterium atom contains a proton and a neutron. The nucleus of a tritium atom contains a proton and two neutrons.

Heavy hydrogen is used for two reasons. First, these isotopes fuse at lower temperatures than regular hydrogen does. Second, they are relatively common. About 1 in 6,500 of the hydrogen atoms in natural water are deuterium atoms. Tritium breaks down rapidly, so very little of it is found in nature. It is made artificially by exposing the metal lithium to fast-moving neutrons created in a nuclear reactor.

If a mixture of deuterium and tritium is made hot enough, some of the deuterium nuclei fuse with tritium nuclei. One deuterium nucleus fuses with one tritium nucleus to produce one helium nucleus. (Helium is the gas that is used to fill party balloons.) When this happens, energy is given off in the form of a fast-moving neutron. This also happens in a hydrogen bomb, but it doesn't have to happen as a huge explosion: in theory, it could happen as slowly as one atom at a time.

Some scientists argue that fusion could be the "energy of the future" because its fuel—heavy hydrogen—contains an enormous amount of energy by weight. A bottle-cap full of heavy hydrogen contains the same amount of energy as twenty tons of coal. Further, using such fuel would be relatively safe. The major by-product is helium, which is harmless. A fusion explosion could not happen because there would not be enough hydrogen, and it would not be not packed together the right way. In fact, keeping a fusion reaction going at all has been difficult for scientists trying to build fusion generators.

Because of the high temperatures needed to keep a fusion reaction going, no container made of any known substance such as steel or titanium can be used as a vessel for the reaction. A fusion reaction would simply melt the container and could not be contained or used. One possible solution is to use magnets to hold the reaction.

Controlled fusion begins with the making of a plasma, a form of gas so hot that the nuclei of all of the atoms have been stripped of their electrons. This leaves each nucleus with a positive electrical charge (usually the positive charge of each nucleus is balanced out by the atom's negative electrons). Because a plasma is charged, it can be held in place, or "bottled," by magnetic fields. Ordinary solid materials cannot be used because the plasma is too hot; even steel would simply turn into a gas at such temperatures, like boiling water turns to steam. The magnetic bottle method was developed early on by the Russian scientists who invented the device called a *tokamak*. "Tokomak" is short for "toroidal magnetic chamber" in Russian, where "toroidal" means doughnut-shaped.

A tokomak is a steel chamber shaped like a hollow doughnut. Plasma is held inside the doughnut by magnetic fields and heated. When it is hot enough, fusion begins. The magnetic fields are supposed to keep the plasma from touching the inside walls of the reactor. So far, the main problem with tokamaks is that the plasma leaks out of the magnetic fields when the fusion reaction begins so that the reaction can be kept going for only a few seconds. Only if this problem can be overcome can tokamak containers house useable fusion reactions. Several large tokomaks have been built, but none has produced as much energy as it takes to run.

On June 28, 2005, six partners (China, Japan, South Korea, Russia, the United States, and the European Union) agreed on a site for a tokomak to be called the International Thermonuclear Experimental Reactor (ITER). ITER will be built in Cadarache, north of Marseille, France. This is a multi-billion-dollar project designed to make possible experiments that the sponsors hope will lead to a greater understanding of fusion reactions and eventually to electricity-producing fusion power plants. In December 2005, the ITER site was prepared for construction of the reactor. Its designers currently plan for operation to begin in 2016.

Another way of keeping plasma hot enough for fusion to happen is "inertial confinement." This uses powerful laser beams to blast a tiny pellet of hydrogen fuel from all sides at once, turning it into hot plasma before it can expand and cool. While this method has worked for experimental purposes, scientists doubt whether it can ever be a feasible source of commercial power.

To be a useful power source, a fusion reactor would not only have to make more energy than it uses, but it would have to make that energy more cheaply than other sources of energy can be made. But there seems to be only a small chance that fusion can be made to produce large amounts of power, at any price, for many years to come.

Cold fusion

Because it is so hard to control the star-like temperatures needed for "hot" fusion, some scientists have looked for ways to make fusion happen at low temperatures. This is sometimes called "cold" fusion, a term coined in 1986 by Dr. Paul Palmer of Utah's Brigham Young University. As with zero point energy, all physicists agree that certain forms of cold fusion do happen, but most do not think that cold fusion can ever be a practical source of energy.

The history of cold fusion began in the nineteenth century, when scientists recognized the unique ability of the metals palladium and titanium to absorb hydrogen, much as sponges absorb water. In the twentieth century, scientists thought that these elements might be able to hold deuterium atoms so close together that a fusion reaction would result even at low temperatures. Later, two German scientists claimed to have performed an experiment using palladium that transformed hydrogen into helium at room temperature. However, they later took back their claim, admitting that the helium had probably come from the surrounding air.

The Pons-Fleischmann announcement

In the following decades, a few scientists around the world continued to experiment with ways to produce fusion at low temperatures. None succeeded, but by the 1980s, after the energy shortages of the 1970s, a few scientists still worked on the premise that cold fusion held out hope for a future of clean, safe, abundant energy. In 1984, Stanley Pons of the University of Utah and Martin Fleischmann from England's University of Southampton began conducting cold fusion experiments at the University of Utah. On March 23, 1989, Pons and Fleischmann held a press conference at which they made an announcement that startled the world. They claimed that they had successfully carried out a cold fusion experiment that produced excess heat that could be explained only by a fusion reaction, not by chemical processes (such as metal combining with oxygen). At long last, the dream of being able to produce energy on a commercial scale from a bucket of water seemed to be just around the corner.

In their experiment, Pons and Fleischmann used a double-walled vacuum flask to reduce heat conduction. They filled the flask with "heavy water," water made with the deuterium isotope of hydrogen replacing ordinary hydrogen (the "H" in the chemical formula for water, H_2O). They inserted a piece of palladium metal in the heavy water and applied an electrical current. According to their results, nothing happened for a period of weeks. The energy input and energy output of the system were steady, and the temperature of the water stayed at 86°F (30°C). Then the temperature suddenly rose to 122°F (50°C), without any increase in the input power. The water remained at that temperature for two days before

"Cold fusion" palladium and platinum electrodes, part of a French experiment to investigate the results of Fleischmann & Pons, who claimed to have created sustained cold fusion energy production in a simple electrolytic cell.
Philippe Plailly/Photo Researchers, Inc.

decreasing again. This happened more than once. During these power bursts, the energy output was about twenty times greater than the energy input.

Because of the simplicity of the Pons-Fleischmann design, groups of scientists around the world attempted to duplicate their results. For weeks, the topic of cold fusion was on the front pages of newspapers. Some scientists initially reported that they were able to duplicate the Utah experiments, while others failed. What resulted was a mix of claims, theories, explanations, accusations, and arguments that the press dubbed "fusion confusion."

Since 1989, many scientists claim to have produced cold fusion. In some experiments, excess heat is generated. The expected by-products of cold fusion—neutrons, tritium, and charged particles—have been reported. Other laboratories have found the production of an isotope of helium, another potential by-product of fusion. They have also reported isotopes of such elements as silver and rhodium, again suggesting that something is happening at an atomic level.

To a nonscientist, it all sounds pretty convincing. Yet most scientists do not accept that cold fusion has been achieved. There is, to begin with, no theory that would explain it. The minority who argue for cold fusion point out that even though science cannot explain cold fusion, that does not prove that it is not real. They note, for example, that when superconductivity (the flow of electricity through very cold metals with zero loss) was first discovered in the early twentieth century, there was no theory to

Demonstration of magnetic levitation of one of the new high-temperature superconductors yttrium-barium-copper oxide (Y-Ba2-Cu3-O7-x). The small, cylindrical magnet floats freely above a nitrogen-cooled, cylindrical specimen of a superconducting ceramic. The glowing vapor is from liquid nitrogen, which maintains the ceramic within its superconducting temperature range. *David Parker/Photo Researchers, Inc.*

explain it until decades later. This is true, but in the case of cold fusion, the observations themselves are in doubt. The case for cold fusion is not as certain as a mere list of all the positive reports makes it sound.

First, there have also been many failed cold-fusion experiments. Second, the production of energy by a system is not proof that nuclear reactions are happening; chemical reactions could be supplying the energy. Third, the production of "excess heat" by a system—often reported by scientists working with cold fusion setups—does not necessarily mean that more energy is coming out of the system *over the lifetime of the experiment* than goes into it. Fourth, there are many possible sources of measurement error when looking for fusion by-products. Extra helium, for example, may come from the air; silver or rhodium (supposedly detected in extremely small amounts) may come from contaminated instruments; neutrons may come from cosmic rays or radioactive elements such as uranium.

A supercooled superconductor creates magnetic levitation, as well as steam, due to its low temperature. *Charles O'Rear/Corbis.*

As of early 2006, seventeen years after the Pons-Fleischmann announcement, there was still no widely accepted proof that nuclear fusion is happening in the devices built by cold-fusion researchers. The scientific community as a whole has not been convinced that cold fusion is real. That is, they are not convinced that any kind of cold fusion that produces more energy than goes into it is real. There is agreement among physicists that energy-consuming forms of cold fusion do exist. In particular, the phenomenon called "muon-catalyzed fusion" is well-established. Muons are particles that can briefly substitute for electrons in atoms. When they do this they shield the atomic nuclei from each other, reducing the electrical force that keeps them apart and so allowing them to be fused by lower-velocity collisions (cooler temperatures). Muons, however, have a limited lifetime—about 2.2 millionths of a second—and more energy is needed to produce them than they can release through fusion.

In the 1990s, the U.S. Department of Energy suspended funding for cold fusion research. In 2004 it conducted a study in which it concluded that research since 1989 had produced nothing new of substance. Japan continues to fund cold fusion research.

Sonofusion

Claims of another kind of "desktop" fusion (fusion that can be produced by inexpensive, simple equipment rather than multi-billion dollar tokomaks) surfaced in 2002. Physicist Rusi Taleyarkhan of Purdue University published a study claiming to have produced fusion using sonoluminescence. Sonoluminescence—the word means, literally, "sound-light"—occurs in some liquids when they are hit by intense sound waves. Tiny, short-lived bubbles appear in the liquid and then collapse. When each bubble collapses, very high temperatures and pressures occur inside it and a tiny flash of light is given off. Temperatures of thousands of degrees are generated in these collapsing bubbles, but physicists are not sure just how hot they are. If the temperature were high enough, it could cause fusion. Most physicists however currently argue that temperatures can not reach this high level.

Dr. Taleyarkhan ran his first experiments at the Oak Ridge National Laboratory in Tennessee, a laboratory owned by the U.S. government. He used a liquid chemical called acetone. The normal hydrogen atoms in the acetone that Taleyarkhan used had been replaced with atoms of deuterium, one of the heavy forms of hydrogen. He hoped that super-high temperatures in collapsing sonoluminescence bubbles would make the deuterium atoms fuse. To see whether fusion was really happening, he placed detectors around his acetone setup to count fast-moving neutrons. Neutrons would prove that fusion was occurring. Taleyarkhan believed that he counted enough neutrons to prove the presence of fusion.

Taleyarkhan's work was real science, but that doesn't mean it couldn't be wrong. Some other scientists criticized the details of his work. For example, fusion was not the only possible source of the neutrons that Taleyarkhan was measuring; he was shooting neutrons at the acetone to make bubbles form faster. Therefore, to detect fusion, Taleyarkhan had to measure not just whether there were **any** neutrons coming out of the experimental setup, but whether there were **extra** neutrons coming out—a much trickier problem.

Much as with cold fusion, hopes run high for sonofusion. But as of early 2006, no one had been able to duplicate Taleyarkhan's results. However, in early 2006 he announced that he was about to publish new results in the journal *Physics Review Letters*, an important science journal. Most physicists argue that Taleyarkhan is making an honest mistake in his experiments. The scientific process of presenting evidence and testing new ideas will eventually show whether he is correct.

SOLAR POWER SATELLITES

Solar cells or photovoltaic cells, devices that turn sunlight directly into electricity, work best in outer space. The sun is brighter there because there is no air to block any light, and solar cells can be stationed outside the Earth's shadow so they see the sun all the time. In fact, solar cells were first used, in the 1950s, to power space satellites. Some people have argued that we should use solar cells in space to generate power for the Earth. They say that we should build large arrays of solar cells in orbit around the Earth—solar power satellites.

But there is a problem: it is impossible to run power lines from a satellite to the Earth. Any wire or cable would snap under its own weight long before it was long enough to reach from the Earth's surface into space. Therefore, supporters of solar power satellites propose to beam the power to Earth in the form of radio waves. The kind of radio waves that would be used are "microwaves", the same kind that are used to cook food in microwave ovens.

The system would look like this: a large, flat array of solar cells would orbit the Earth at a height of about 22,000 miles (36,000 km). At that height, it takes a satellite 24 hours to circle the Earth. Since the Earth is spinning once every 24 hours, a satellite at that height (circling in the same direction as the Earth is turning) looks from the ground like it is standing still in the sky (geostationary, that is, remaining above the same point over the ground). Satellites of this kind are used to broadcast satellite TV signals. Also, a satellite that far from the Earth can be positioned so that the Earth's shadow never falls across it and breaks the supply of sunlight.

This giant array of solar cells would make electricity, turn it into radio waves, and beam the radio waves at Earth. A large antenna on the ground would pick up the radio waves and turn them into electricity again. The power would then be transmitted to users through power lines, just as power from ordinary generating plants is.

There are no basic scientific problems with this idea: everything about it uses machines that we already know how to make. The great problem is cost. A solar-cell array and microwave radio transmitter of the size needed would weigh many tons. The cost of launching all that machinery with rockets would be huge—far greater than the cost of building solar power stations, windmills, and other sources of renewable power right here on Earth. Although there is nothing basically wrong with the idea of solar power stations, they would be difficult to finance and build. Only wealthy and technologically advanced governments could currently fund such an effort. Only Japan has announced intentions to at least explore the possibility, but not until 2040.

NO MAGIC BULLETS

Hot fusion and solar power satellites are based on solid science, but there seems to be no current way to make them practical or affordable, at least for the foreseeable future. Cold fusion, sono-fusion, and zero point energy, on the other hand, are based on scientific claims that most scientists currently reject. And perpetual motion is a complete fake that is not possible because of the well-tested laws of thermodynamics. Accordingly, there is probably not going to be any near-term "magic bullet" for our energy problems. We already know what tools we have to choose from: fossil fuels, nuclear power, and renewable energy sources such as the wind and sun, geothermal power, biofuels, wave or tide energy, and hydro-electric power.

There is intense disagreement in our society over what the right energy choices are that are both possible and affordable. For example, some people claim that it would be madness to not develop nuclear power on a huge scale, and others say it would be a disaster to do so. Some say that renewable energy can supply all our needs, and others that such energy sources can not meet increasing energy demands. There is no easy answer to the energy problem; even the best answers developed in the near future may be complicated, dangerous, and expensive. However, one thing is certain: **all** ways of making energy harm the Earth to some extent. Therefore, no matter where our energy comes from, we should not waste it. Living a more energy-efficient life is as easy as reaching out to turn off the nearest unneeded light.

Even as scientists and engineers are working on more efficient refrigerators, cars, computers, lights, and other devices, we can all save a significant amount of energy just by turning off lights, computers, and other devices whenever we aren't using them. Over time, we all make many choices about how much energy to use and how to use it. A more energy-efficient world is a world that is easier to supply with energy, whatever the source.

■■■

For More Information

Books

Close, Frank E. *Too Hot to Handle: The Race for Cold Fusion.* Princeton, NJ: Princeton University Press, 1991.

Cook, Nick.*The Hunt for Zero Point: Inside the Classified World of Antigravity Technology.* New York: Broadway Books, 2003.

Ord-Hume, Arthur W. J. G. *Perpetual Motion: The History of an Obsession.* New York: St. Martin's Press, 1980.

Web Sites

Fukada, Takahiro. "Japan Plans To Launch Solar Power Station In Space By 2040." *SpaceDaily.com,* January 1, 2001. Available at http://www.spacedaily.com/news/ssp-01a.html (accessed on February 12, 2006).

Lovins, Amory. "Mighty Mice: The most powerful force resisting new nuclear may be a legion of small, fast and simple microgeneration and efficiency projects." *Nuclear Engineering International,* December 2005. Available at http://www.rmi.org/images/other/Energy/E05-15_MightyMice.pdf (accessed on February 12, 2006).

U.S. Department of Energy. *Report of the Review of Low Energy Nuclear Reactions.* Washington, DC: Department of Energy, December 1, 2004. http://lenr-canr.org/acrobat/DOEreportofth.pdf, accessed on February 12, 2006).

Yam, Philip. "Exploiting Zero-Point Energy." *Scientific American,* December 1997. Available from http://www.padrak.com/ine/ZPESCIAM.html (accessed on August 2, 2005).

■■■
Where to Learn More

BOOKS

Angelo, Joseph A. *Nuclear Technology*. Westport, CT: Greenwood Press, 2004.

Avery, William H., and Chih Wu. *Renewable Energy from the Ocean*. New York: Oxford University Press, 1994.

Berinstein, Paula. *Alternative Energy: Facts, Statistics, and Issues*. Phoenix, AZ: Oryx Press, 2001.

Boyle, Godfrey. *Renewable Energy, 2nd ed.* New York: Oxford University Press, 2004.

Buckley, Shawn. *Sun Up to Sun Down: Understanding Solar Energy*. New York: McGraw-Hill, 1979.

Burton, Tony, David Sharpe, Nick Jenkins, and Ervin Bossanyi. *Wind Energy Handbook*. New York: Wiley, 2001.

Carter, Dan M., and Jon Halle. *How to Make Biodiesel*. Winslow, Bucks, UK: Low-Impact Living Initiative (Lili), 2005.

Cataldi, Raffaele, ed. *Stories from a Heated Earth: Our Geothermal Heritage*. Davis, CA: Geothermal Resources Council, 1999.

Close, Frank E. *Too Hot to Handle: The Race for Cold Fusion*. Princeton, NJ: Princeton University Press, 1991.

Cook, Nick.*The Hunt for Zero Point: Inside the Classified World of Antigravity Technology*. New York: Broadway Books, 2003.

Cuff, David J., and William J. Young. *The United States Energy Atlas,* 2nd ed. New York: Macmillan, 1986.

Dickson, Mary H., and Mario Fanelli, eds. *Geothermal Energy: Utilization and Technology*. London: Earthscan Publications, 2005.

Domenici, Peter V. *A Brighter Tomorrow : Fulfilling the Promise of Nuclear Energy*. Lanham, MD: Rowman and Littlefield, 2004.

Ewing, Rex. *Hydrogen: Hot Cool Science—Journey to a World of the Hydrogen Energy and Fuel Cells at the Wassterstoff Farm.* Masonville, CO: Pixyjack Press, 2004.

Freese, Barbara. *Coal: A Human History.* New York: Perseus, 2003.

Frej, Anne B. *Green Office Buildings: A Practical Guide to Development.* Washington, DC: Urban Land Institute, 2005.

Gelbspan, Ross. *Boiling Point: How Politicians, Big Oil and Coal, Journalists and Activists Are Fueling the Climate Crisis.* New York: Basic Books, 2004.

Geothermal Development in the Pacific Rim. Davis, CA: Geothermal Resources Council, 1996.

Graham, Ian. *Geothermal and Bio-Energy.* Fort Bragg, CA: Raintree, 1999.

Heaberlin, Scott W. *A Case for Nuclear-Generated Electricity: (Or Why I Think Nuclear Power Is Cool and Why It Is Important That You Think So Too).* Columbus, OH: Battelle Press, 2003.

Howes, Ruth, and Anthony Fainberg. *The Energy Sourcebook: A Guide to Technology, Resources and Policy.* College Park, MD: American Institute of Physics, 1991.

Husain, Iqbal. *Electric and Hybrid Vehicles: Design Fundamentals.* Boca Raton, FL: CRC Press, 2003.

Hyde, Richard. *Climate Responsive Design.* London: Taylor and Francis, 2000.

Kaku, Michio, and Jennifer Trainer, eds. *Nuclear Power: Both Sides.* New York: Norton, 1983.

Kibert, Charles J. *Sustainable Construction: Green Building Design and Delivery.* New York: Wiley, 2005.

Leffler, William L. *Petroleum Refining in Nontechnical Language.* Tulsa, OK: Pennwell Books, 2000.

Lusted, Marcia, and Greg Lusted. *A Nuclear Power Plant.* San Diego, CA: Lucent Books, 2004.

Manwell, J. F., J. G. McGowan, and A. L. Rogers. *Wind Energy Explained.* New York: Wiley, 2002.

McDaniels, David K. *The Sun.* 2nd ed. New York: John Wiley & Sons, 1984.

Morris, Robert C. *The Environmental Case for Nuclear Power.* St. Paul, MN: Paragon House, 2000.

National Renewable Energy Laboratory, U.S. Department of Energy. *Wind Energy Information Guide.* Honolulu, HI: University Press of the Pacific, 2005.

Ord-Hume, Arthur W. J. G. *Perpetual Motion: The History of an Obsession*. New York: St. Martin's Press, 1980.

Pahl, Greg. *Biodiesel: Growing a New Energy Economy*. Brattleboro, VT: Chelsea Green Publishing Company, 2005.

Rifkin, Jeremy. *The Hydrogen Economy*. New York: Tarcher/Putnam, 2002.

Romm, Joseph J. *The Hype of Hydrogen: Fact and Fiction in the Race to Save the Climate*. Washington, DC: Island Press, 2004.

Seaborg, Glenn T. *Peaceful Uses of Nuclear Energy*. Honolulu, HI: University Press of the Pacific, 2005.

Tickell, Joshua. *From the Fryer to the Fuel Tank: The Complete Guide to Using Vegetable Oil as an Alternative Fuel*. Covington, LA: Tickell Energy Consultants, 2000.

Wohltez, Kenneth, and Grant Keiken. *Volcanology and Geothermal Energy*. Berkeley: University of California Press, 1992.

Wulfinghoff, Donald R. *Energy Efficiency Manual: For Everyone Who Uses Energy, Pays for Utilities, Designs and Builds, Is Interested in Energy Conservation and the Environment*. Wheaton, MD: Energy Institute Press, 2000.

PERIODICALS

Anderson, Heidi. "Environmental Drawbacks of Renewable Energy: Are They Real or Exaggerated?" *Environmental Science and Engineering* (January 2001).

Behar, Michael. "Warning: The Hydrogen Economy May Be More Distant Than It Appears." *Popular Mechanics* (January 1, 2005): 64.

Brown, Kathryn. "Invisible Energy." *Discover* (October 1999): 36.

Burns, Lawrence C., J. Byron McCormick, and Christopher E. Borroni-Bird. "Vehicles of Change." *Scientific American* (October 2002): 64-73.

Corcoran, Elizabeth. "Bright Ideas." *Forbes* (November 24, 2003): 222.

Dixon, Chris. "Shortages Stifle a Boom Time for the Solar Industry." *New York Times* (August 5, 2005): A1.

Feldman, William. "Lighting the Way: To Increased Energy .Efficiency." *Journal of Property Management* (May 1, 2001): 70.

Freeman, Kris. "Tidal Turbines: Wave of the Future?" *Environmental Health Sciences* (January 1, 2004): 26.

Graber, Cynthia. "Building the Hydrogen Boom." *OnEarth* (Spring 2005): 6.

Grant, Paul. "Hydrogen Lifts Off—with a Heavy Load." *Nature* (July 10, 2003): 129-130.

Guteral, Fred, and Andrew Romano. "Power People." *Newsweek* (September 20, 2004): 32.

Hakim, Danny. "George Jetson, Meet the Sequel." *New York Times* (January 9, 2005): section 3, p. 1.

Lemley, Brad. "Lovin' Hydrogen." *Discover* (November 2001): 53-57, 86.

Libby, Brian. "Beyond the Bulbs: In Praise of Natural Light." *New York Times* (June 17, 2003): F5.

Linde, Paul. "Windmills: From Jiddah to Yorkshire." *Saudi Aramco World* (January/February 1980). This article can also be found online at http://www.saudiaramcoworld.com/issue/198001/windmills-from.jiddah.to.yorkshire.htm.

Lizza, Ryan. "The Nation: The Hydrogen Economy; A Green Car That the Energy Industry Loves." *New York Times* (February 2, 2003): section 4, p. 3.

McAlister, Roy. "Tapping Energy from Solar Hydrogen." *World and I* (February 1999): 164.

Motavalli, Jim. "Watt's the Story? Energy-Efficient Lighting Comes of Age." *E* (September 1, 2003): 54.

Muller, Joann, and Jonathan Fahey. "Hydrogen Man." *Forbes* (December 27, 2004): 46.

Nowak, Rachel. "Power Tower." *New Scientist* (July 31, 2004): 42.

Parfit, Michael. "Future Power: Where Will the World Get Its Next Energy Fix?" *National Geographic* (August 2005): 2–31.

Pearce, Fred. "Power of the Midday Sun." *New Scientist* (April 10, 2004): 26.

Perlin, John. "Soaring with the Sun." *World and I* (August 1999): 166.

Port, Otis. "Hydrogen Cars Are Almost Here, but . . . There Are Still Serious Problems to Solve, Such As: Where Will Drivers Fuel Up?" *Business Week* (January 24, 2005): 56.

Provey, Joe. "The Sun Also Rises." *Popular Mechanics* (September 2002): 92.

Service, Robert F. "The Hydrogen Backlash." *Science* (August 13, 2004): 958-961.

"Stirrings in the Corn Fields." *The Economist* (May 12, 2005).

Terrell, Kenneth. "Running on Fumes." *U.S. News & World Report* (April 29, 2002): 58.

Tompkins, Joshua. "Dishing Out Real Power." *Popular Science* (February 1, 2005): 31.

Valenti, Michael. "Storing Hydroelectricity to Meet Peak-Hour Demand." *Mechanical Engineering* (April 1, 1992): 46.

Wald, Matthew L. "Questions about a Hydrogen Economy." *Scientific American* (May 2004): 66.

Westrup, Hugh. "Cool Fuel: Will Hydrogen Cure the Country's Addiction to Fossil Fuels?" *Current Science* (November 7, 2003): 10.

Westrup, Hugh. "What a Gas!" *Current Science* (April 6, 2001): 10.

WEB SITES

"Alternative Fuels." U.S. Department of Energy Alternative Fuels Data Center. http://www.eere.energy.gov/afdc/altfuel/altfuels.html (accessed on July 20, 2005).

"Alternative Fuels Data Center." U.S. Department of Energy: Energy Efficiency and Renewable Energy. http://www.eere.energy.gov/afdc/altfuel/p-series.html (accessed on July 11, 2005).

American Council for an Energy-Efficient Economy. http://aceee.org/ (accessed on July 27, 2005).

The American Solar Energy Society. http://www.ases.org/ (accessed on September 1, 2005).

Biodiesel Community. http://www.biodieselcommunity.org/ (accessed on July 27, 2005).

"Bioenergy." Natural Resources Canada. http://www.canren.gc.ca/tech_appl/index.asp?CaId=2&PgId=62 (accessed on July 29, 2005).

"Biofuels." Journey to Forever. http://journeytoforever.org/biofuel.html (accessed on July 13, 2005).

"Biogas Study." Schatz Energy Research Center. http://www.humboldt.edu/~serc/biogas.html (accessed on July 15, 2005).

"Black Lung." United Mine Workers of America. http://www.umwa.org/blacklung/blacklung.shtml (accessed on July 20, 2005).

"Classroom Energy!" American Petroleum Institute. http://www.classroom-energy.org (accessed on July 20, 2005).

"Clean Energy Basics: About Solar Energy." National Renewable Energy Laboratory. http://www.nrel.gov/clean_energy/solar.html (accessed on August 25, 2005).

"A Complete Guide to Composting." Compost Guide. http://www.compostguide.com/ (accessed on July 25, 2005).

"Conserval Engineering, Inc." American Institute of Architects. http://www.solarwall.com/ (accessed on September 1, 2005).

"The Discovery of Fission." Center for History of Physics. http://www.aip.org/history/mod/fission/fission1/01.html (accessed on December 17, 2005).

"Driving for the Future." California Fuel Cell Partnership. http://www.cafcp.org (accessed on August 8, 2005).

"Driving and Maintaining Your Vehicle." Natural Resources Canada. http://oee.nrcan.gc.ca/transportation/personal/driving/autosmart-maintenance.cfm?attr=11 (accessed on September 28, 2005).

"Ecological Footprint Quiz." Earth Day Network. http://www.earthday.net/footprint/index.asp (accessed on February 6, 2006).

"Energy Efficiency and Renewable Energy." U.S. Department of Energy. http://www.eere.energy.gov (accessed on September 28, 2005).

"Ethanol: Fuel for Clean Air." Minnesota Department of Agriculture. http://www.mda.state.mn.us/ethanol/ (accessed on July 14, 2005).

"Florida Solar Energy Center." University of Central Florida. http://www.fsec.ucf.edu (accessed on September 1, 2005).

"Fueleconomy.gov." United States Department of Energy. http://www.fueleconomy.gov/feg/ (accessed on July 27, 2005).

Fukada, Takahiro. "Japan Plans To Launch Solar Power Station In Space By 2040." *SpaceDaily.com,* Jan. 1, 2001. Available at http://www.spacedaily.com/news/ssp-01a.html (accessed Feb. 12, 2006).

"Geo-Heat Center." Oregon Institute of Technology. http://geoheat.oit.edu. (accessed on July 19, 2005).

"Geothermal Energy." World Bank. http://www.worldbank.org/html/fpd/energy/geothermal. (accessed on July 19, 2005).

Geothermal Resources Council. http://www.geothermal.org. (accessed on August 4, 2005).

"Geothermal Technologies Program." U.S. Department of Energy: Energy Efficiency and Renewable Energy. http://www.eere.energy.gov/geothermal. (accessed on July 22, 2005).

"Green Building Basics." California Home. http://www.ciwmb.ca.gov/GreenBuilding/Basics.htm (accessed on September 28, 2005).

"Guided Tour on Wind Energy." Danish Wind Industry Association. http://www.windpower.org/en/tour.htm (accessed on July 25, 2005).

"How the BMW H2R Works." How Stuff Works. http://auto.howstuffworks.com/bmw-h2r.htm (accessed on August 8, 2005).

"Hydrogen, Fuel Cells & Infrastructure Technologies Program." U.S. Department of Energy Energy Efficiency and Renewable Energy. http://www.eere.energy.gov/hydrogenandfuelcells/ (accessed on August 8, 2005).

"Hydrogen Internal Combustion." Ford Motor Company. http://www.ford.com/en/innovation/engineFuelTechnology/hydrogenInternalCombustion.htm (accessed on August 8, 2005).

"Incandescent, Fluorescent, Halogen, and Compact Fluorescent." California Energy Commission. http://www.consumerenergycenter.org/homeandwork/homes/inside/lighting/bulbs.html (accessed on September 28, 2005).

"Introduction to Green Building." Green Roundtable. http://www.greenroundtable.org/pdfs/Intro-To-Green-Building.pdf (accessed on September 28, 2005).

Lovins, Amory. "Mighty Mice: The most powerful force resisting new nuclear may be a legion of small, fast and simple micro-generation and efficiency projects." *Nuclear Engineering International,* Dec. 2005. Available at http://www.rmi.org/images/other/Energy/E05-15_MightyMice.pdf (accessed Feb. 12, 2006).

Nice, Karim. "How Hybrid Cars Work." Howstuffworks.com. http://auto.howstuffworks.com/hybrid-car.htm (accessed on September 28, 2005).

"Nuclear Terrorism—How to Prevent It." Nuclear Control Institute. http://www.nci.org/nuketerror.htm (accessed on December 17, 2005).

"Oil Spill Facts: Questions and Answers." *Exxon Valdez* Oil Spill Trustee Council. http://www.evostc.state.ak.us/facts/qanda.html (accessed on July 20, 2005).

O'Mara, Katrina, and Mark Rayner. "Tidal Power Systems." http://reslab.com.au/resfiles/tidal/text.html (accessed on September 13, 2005).

"Photos of El Paso Solar Pond." University of Texas at El Paso. http://www.solarpond.utep.edu/page1.htm (accessed on August 25, 2005).

"The Plain English Guide to the Clean Air Act." U.S. Environmental Protection Agency. http://www.epa.gov/air/oaqps/peg_caa/pegcaain.html (accessed on July 20, 2005).

"Reinventing the Automobile with Fuel Cell Technology." General Motors Company. http://www.gm.com/company/gmability/adv_tech/400_fcv/ (accessed on August 8, 2005).

"Safety of Nuclear Power." Uranium Information Centre, Ltd. http://www.uic.com.au/nip14.htm (accessed on December 17, 2005).

"Solar Energy for Your Home." Solar Energy Society of Canada Inc. http://www.solarenergysociety.ca/2003/home.asp (accessed on August 25, 2005).

The Solar Guide. http://www.thesolarguide.com (accessed on September 1, 2005).

"Solar Ponds for Trapping Solar Energy." United National Environmental Programme. http://edugreen.teri.res.in/explore/renew/pond.htm (accessed on August 25, 2005).

"Thermal Mass and R-value: Making Sense of a Confusing Issue." BuildingGreen.com. http://buildggreen.com/auth/article.cfm?fileName=070401a.xml (accessed on September 28, 2005).

"Tidal Power." University of Strathclyde. http://www.esru.strath.ac.uk/EandE/Web_sites/01-02/RE_info/Tidal%20Power.htm (accessed on September 13, 2005).

U.S. Department of Energy. *Report of the Review of Low Energy Nuclear Reactions.* Washington, DC: Department of Energy, Dec. 1, 2004. http://lenr-canr.org/acrobat/DOEreportofth.pdf, accessed Feb. 12, 2006.

Vega, L. A. "Ocean Thermal Energy Conversion (OTEC)." http://www.hawaii.gov/dbedt/ert/otec/index.html (accessed on September 13, 2005).

Venetoulis, Jason, Dahlia Chazan, and Christopher Gaudet. "Ecological Footprint of Nations: 2004." Redefining Progress. http://www.rprogress.org/newpubs/2004/footprintnations2004.pdf (accessed on February 8, 2006.)

Weiss, Peter. "Oceans of Electricity." *Science News Online* (April 14, 2001). http://www.science news.org/articles/20010414/bob12.asp (accessed on September 13, 2005).

"What Is Uranium? How Does It Work?" World Nuclear Association. http://www.world-nuclear.org/education/uran.htm (accessed on December 17, 2005).

"Wind Energy Tutorial." American Wind Energy Association. http://www.awea.org/faq/index.html (accessed on July 25, 2005).

Yam, Philip. "Exploiting Zero-Point Energy." *Scientific American*, December 1997. Available from http://www.padrak.com/ine/ZPESCIAM.html (accessed on August 2, 2005).

OTHER SOURCES

World Spaceflight News. *21st Century Complete Guide to Hydrogen Power Energy and Fuel Cell Cars: FreedomCAR Plans, Automotive Technology for Hydrogen Fuel Cells, Hydrogen Production, Storage, Safety Standards, Energy Department, DOD, and NASA Research.* Progressive Management, 2003.

Index

Italic type indicates volume number; boldface type indicates entries and their pages; (ill.) indicates illustrations.

A

Abraham, Spencer, *2:* 143
AC (Alternating current), *2:* 243
Acetone, *3:* 406
Acetylene, *1:* 48
Acid rain, *1:* 14, 15-16, 41
Active solar systems, *2:* 218
Adams, William G., *2:* 212, 214
Adobe, *2:* 210, *3:* 349
Aerodynamics, *3:* 368-69
Aeromotor Company, *3:* 311-12
Afghanistan
 hydropower in, *3:* 277
 windmills in, *3:* 306
Agriculture
 for biofuels, *1:* 66-67
 geothermal energy for, *1:* 115-18
 hydroelectric dams and, *3:* 286
 open field, *1:* 115-16
 See also Farms
Air conditioning
 natural gas for, *1:* 35
 ocean thermal energy conversion for, *3:* 292
 See also Cooling
Air pollution, *1:* 17 (ill.)
 from biofuels, *1:* 65

from coal, *1:* 20, 40-42, 43, *2:* 202, *3:* 340
from coal gasification, *1:* 45
from ethanol, *1:* 91
from exhaust emissions, *1:* 6, 12, 93, *2:* 142-43
from fossil fuels, *1:* 12-15, 20
from gasoline, *1:* 28
indoor, *1:* 14-15, 16, *3:* 347, 348
from methane, *1:* 86-87
from methanol, *1:* 51-52
MTBE and, *1:* 53
from natural gas, *1:* 37
nuclear power plants and, *2:* 202
particulate matter, *1:* 12-13, 14, 16
from P-Series fuels, *1:* 93
sources of, *1:* 14
Air quality standards, *1:* 124
Airplanes
 history of, *1:* 10
 lift and drag on, *3:* 324-25
 solar-powered, *2:* 219
 weight on, *3:* 373
 zero point energy for, *3:* 395
Airships, *2:* 137-39, 140 (ill.), 141, 142 (ill.)
Akron (Airship), *2:* 138
Alberta, Canada oil sands, *1:* 25

Alcohol fuels, 1:87-92
Al-Dimashqi, *3:* 308
Alexander the Great, *1:* 41
Algae
 biodiesel from, *1:* 75
 hydrogen from, *2:* 149
Alkaline fuel cells, *2:* 145-46
al-Qaeda, *2:* 199
Alternating current (AC), *2:* 243
Alternative energy, *3:* 380-84
 See also Renewable energy; specific types of alternative energy
Aluminum cans, recycled, *3:* 376
American Society of Civil Engineers, *3:* 283
American Wind Energy Association, *3:* 319
Ammonia
 biogas, *1:* 84
 for ocean thermal energy conversion, *3:* 290, 293
 in solar collectors, *2:* 215
Amorphous photovoltaic cells, *2:* 237-38
Anaerobic digestion technology, *1:* 84
Anasazi Indians, *3:* 343
Anemometers, *3:* 327
Animal waste. *See* Dung; Manure

Animal-based food products, 3: 376-77
Anthracite, 1: 39
Antifreeze, 1: 90
Anti-knocking agents, 1: 6, 27, 90
Apollo (God), 2: 209-10
Apollo spacecraft, 2: 139-40
Appliances, energy efficient, 3: 360-64
Aquaculture, geothermal energy for, 1: 110, 118-21
Aquatic plants, 1: 120
Aqueducts, 3: 261-62
Argentina, geothermal energy in, 1: 116
Argon, 1: 34
Aristotle
    passive solar design by, 3: 342
    on vacuums, 3: 390, 391
Arizona Public Service Company, 2: 244
Arkwright, Richard, 3: 262, 265
Arsenic, 1: 41
Asphalt, 1: 27
Assist hybrid vehicles, 3: 370
Associated natural gas, 1: 33
Aswan Dam, 3: 286
Atmospheric hydrogen, 2: 162
Atomic bombs, 2: 176-79, 200-201, 3: 398-99
Atomic energy. *See* Nuclear energy
Atomic Energy Commission, 2: 179, 3: 384
Atomic numbers, 2: 170, 172
Atomic weight, 2: 172, 173
Atoms, 2: 169-70
Atoms for Peace program, 2: 178-79
Augers, 3: 307-8
Australia
    solar ponds in, 2: 252
    solar towers in, 2: 254
    tidal power in, 3: 297
Austria, hydroelectricity in, 3: 272
Automatic transmission, 3: 372
Automobiles

ethanol for, 1: 60
gas mileage of, 1: 32, 3: 364-65, 371
history of, 1: 10
internal combustion engines for, 1: 4-6, 5 (ill.)
liquefied petroleum gas for, 1: 48-49
methanol for, 1: 51
solar, 2: 232, 239 (ill.)
tips for fuel-efficient driving, 3: 372-74
See also Hybrid vehicles; Hydrogen fuel cell vehicles; Vehicles
Aviation gasoline, 1: 27

B

Bacteria
    for clean coal technology, 1: 43
    for digestion technology, 1: 85
    for hydrogen production, 2: 149
Bagasse, 1: 70, 71 (ill.), 74, 75, 92
Bahia Blanca, Argentina, 1: 116
Balloons
    hot air, 2: 136
    hydrogen, 2: 135 (ill.), 136
Balneology, 1: 110
Barium, 1: 41
Batch heaters. *See* Integral collector storage (ICS) systems
Bath, England, 1: 100
Baths
    hot springs for, 1: 98-100, 101 (ill.), 110
    Roman, 1: 100, 3: 343
    solar heating for, 2: 210-12
Battelle Pacific Northwest Laboratory, 3: 318-19
Batteries
    in hybrid vehicles, 3: 367
    vs. hydrogen fuel cells, 2: 138
    for photovoltaic cells, 2: 237-38
Bay of Fundy, 3: 298
Bears, polar, 2: 211, 211 (ill.)

Becquerel, Alexandre-Edmond, 2: 212
Bell Laboratories, 2: 213
Benz, Karl, 1: 5 (ill.)
Berthelot, Pierre Eugéne Marcelin, 1: 50
Bessler, Johann, 3: 385
Bessler wheel, 3: 385
Binary geothermal power plants, 1: 103, 108-9, 122, 123 (ill.)
Biodiesel, 1: 64 (ill.), 75-80, 79 (ill.)
    benefits and drawbacks of, 1: 64, 77-78
    economic impact of, 1: 67, 78, 79-80
    environmental impact of, 1: 65, 78
    with gasoline, 1: 75
    history of, 1: 59
    issues and problems with, 1: 80
    making your own, 1: 76, 77-78
    petrodiesel with, 1: 77, 78
    production of, 1: 76
    from rapeseed oil, 1: 61, 75
    uses of, 1: 76-77
    vegetable oil for, 1: 63
Bioenergy, 1: 57-95, 68 (ill.)
    benefits and drawbacks of, 1: 63-65
    environmental impact of, 1: 65-67
    history of, 1: 59-61
    technology for, 1: 62-63
    types of, 1: 58-59
    uses of, 1: 61-62
    See also Biogas; Ethanol; Solid biomass; Vegetable oil fuels
Biofuels
    barriers to, 1: 68-69
    benefits and drawbacks of, 1: 63-65
    economic impact of, 1: 67
    with fossil fuels, 1: 61, 65
    fossil fuels for production of, 1: 67

for internal combustion
engines, *1:* 61-62
issues and problems with, *1:* 87
liquid, *1:* 58
P-Series, *1:* 63, 72, **92-94**
societal impact of, *1:* 67-68
sources of, *1:* 57
technology for, *1:* 62-63
types of, *1:* 58-59
See also Biodiesel; Biogas
Biogas, *1:* 59, 86 (ill.)
benefits and drawbacks of,
1: 85
economic impact of, *1:* 87
environmental impact of,
1: 86-87
from garbage, *1:* 72, 85
for hydrogen production,
2: 149
from manure, *1:* 62, 85
pipelines, *1:* 62
uses of, *1:* 85
Biogas, *1:* 84-87
Biomass
definition of, *1:* 57
P-Series fuels from, *1.* 92
solid, *1:* 58, **69-75**
sources of, *1:* 69
BIPV (Building integrated
photovoltaics), *2:* 240
Birds, wind farms and, *3:* 332-33
Bissell, George, *1:* 10
Bituminous coal, *1:* 39
Blanchard, Jean, *2:* 136
Blimps, *2:* 137-39
BMW, *2:* 155-56, 158 (ill.)
Boats, hydrogen fuel cell-
powered, *2:* 146-47
Bohr, Niels, *2:* 173
Boiling water reactor system,
2: 189
Boise, Idaho, *1:* 100, 102
Bombs
atomic, *2:* 176-79, 200-201,
3: 398-99
dirty, *2:* 200-201
nuclear fusion in, *3:* 398-99
Bonneville Power
Administration, *3:* 282
Borax, *1:* 71

Boron, *2:* 188, 237
Bottles, recycled, *3:* 376
Boyle, Robert, *1:* 50, *2:* 134
BP Amoco, *2:* 242
Braking systems, regenerative,
3: 367-68
Brannan, Sam, *1:* 99-100
Brazil, ethanol production in,
1: 60, 60 (ill.), 70
Breadbox heaters. *See* Integral
collector storage (ICS)
systems
British Hydropower
Association, *3:* 277-78
British thermal units (Btu),
1: 39
Brush, Charles F., *3:* 313
Btu (British thermal units),
1: 39
Building integrated
photovoltaics (BIPV),
2: 240
Building materials
green, *3:* 341, **347-52**
recycled, *3:* 348, 351
Buildings
adobe, *2:* 210, *3:* 349
climate-responsive,
3: **341-47**, 343 (ill.)
cob, *3:* 349-50
commercial, *3:* 345,
346-47
earth bag, *3:* 350
light straw, *3:* 349-50
rammed earth, *3:* 350
remodeled, *3:* 352
sick, *3:* 347, 348
straw bale, *3:* 350
See also Passive solar design
Bulk storage systems. *See*
Integral collector storage
(ICS) systems
Bureau of Labor Statistics
(U.S.), *2:* 196
Burning glasses, *2:* 212
Buses, hydrogen fuel cell,
2: 146, 151, 153 (ill.)
Bush, George W., *1:* 41-42
Butane, *1:* 10, 35, 46-50
Butanol, *1:* 87

C
Cadmium, *2:* 174-76
Cadmium telluride, *2:* 237
CAFE (Corporate Average Fuel
Economy), *1:* 32
California
climate-responsive buildings
in, *3:* 346-47
cost of electricity in, *3:* 333
geothermal energy in,
1: 103, 122
hybrid vehicles in, *3:* 369-70
hydrogen filling stations in,
2: 160
hydrogen fuel cells and,
2: 143
MTBE and, *1:* 54
solar towers in, *2:* 253-54
trough systems in, *2:* 246
wind energy in, *3:* 319, 320,
330
California Air Resources Board,
2: 154
California Energy Commission,
1: 52, *3:* 357
California Fuel Cell
Partnership, *2:* 143
California Hydrogen Highway
Network, *2:* 160
Calories, *3:* 268
Caltrans District 7
Headquarters, *3:* 344 (ill.)
Cameras, infrared, *2:* 211
Cameroon, hydrogen cloud in,
2: 150
Camp stoves, *1:* 4
Campfires, *1:* 62
Canada
biofuels in, *1:* 67
biogas in, *1:* 85
hydroelectricity in, *3:* 272,
286, 288-89
hydrogen research in, *2:*
145-46
liquefied petroleum gas
pipelines from, *1:* 47
oil sands of, *1:* 25
tidal power sites in,
3: 298
Cape Cod, *3:* 270

Carbon dioxide
  from biofuels, *1:* 65
  biogas, *1:* 84, 87
  from coal, *2:* 202
  from commercial buildings,
    3: 345
  from fossil fuels, *1:* 17
  from geothermal energy,
    1: 112
  in hydrogen production,
    2: 148, 150, 162
  methane from, *1:* 33
  natural gas and, *1:* 34, 37
  sources of, *1:* 18
Carbon, from biofuels, *1:* 65
Carbon monoxide, *1:* 14-15
  from coal, *1:* 41
  from ethanol, *1:* 91
  from gasoline, *1:* 28
  in hydrogen production,
    2: 148
  from natural gas, *1:* 37
Carbon monoxide poisoning,
    1: 15
Carbon-based filament,
    3: 353-54
Carboniferous period, *1:* 38
Carpools, *3:* 342 (ill.), 372
Cars. *See* Automobiles
Casimir effect, *3:* 394, 395
Casimir, H. B. G., *3:* 394
Catalysts
  for coal gasification, *1:* 46
  for hydrogen production,
    2: 147-48, 165
Catfish farms, *1:* 119-20
Cavendish, Henry, *2:* 134
Cellular telephones, fuel cells
  for, *1:* 52
Cellulose, *1:* 90
Celts, hot springs and, *1:* 100
Cement composite siding, *3:* 348
Central receivers. *See* Solar
  towers
Centralia, Pennsylvania coal
  mine fires, *1:* 41
Challenger (Shuttle), *2:* 140,
  144 (ill.)
Chapin, Daryl, *2:* 213
Charcoal, *1:* 62, 70-71, 74-75

Charles, Jacques, *2:* 136
Chemical synthesis gas, *1:* 45
Chernobyl nuclear power
  plant, *2:* 183, 185 (ill.), 192
  (ill.), 194, 195
China
  biofuels in, *1:* 67
  biogas in, *1:* 87
  coal-burning engines in, *1:* 7
  coal-burning power plants
    in, *1:* 42
  gasoline use in, *3:* 364
  hydroelectric dams in, *3:*
    285 (ill.), 286, 287 (ill.)
  magnetism in, *3:* 387
  petroleum consumption in,
    1: 19
  photovoltaic cells in, *2:* 242
  waterwheels in, *3:* 261
  wind energy in, *3:* 305-6,
    320
The China Syndrome, *2:* 182
  (ill.)
Chu-Tshai, Yehlu, *3:* 306
Claude, Georges, *3:* 267
Clean Air Act
  on coal plants, *1:* 41-42
  diesel engines and, *1:* 6
  on fuel economy, *1:* 32
  hydrogen vehicles and,
    2: 142-43
  on oxygenated gasoline,
    1: 53
  role of, *1:* 15
Clean coal technology, *1:* 43, 45
Clean Urban Transport for
  Europe program, *2:* 159
Clear Skies program, *1:* 41-42
Clerestory windows, *2:* 226
Climate-responsive buildings,
  3: 341-47, 343 (ill.), 344 (ill.)
Closed-cycle systems, *3:* 290-91,
  291-92
Clothes dryers, energy efficient,
  3: 362
CNG (Compressed natural
  gas), *1:* 36
Coal, *1:* 1, 38-46
  air pollution from, *1:* 20,
    40-42, 43, *2:* 202, *3:* 340

clean coal technology, *1:* 43,
  45
conservation of, *3:* 339-41
deposits of, *1:* 3, 38, *3:* 339
economic impact of, *1:* 42
environmental impact of,
  1: 40-42
finding, *1:* 38
formation of, *1:* 2, 38
history of, *1:* 8-9
issues and problems with,
  1: 43
liquefaction, *1:* 43
processing, *1:* 39-40
pulverization, *1:* 43
societal impact of, *1:* 42-43
vs. solid biomass fuels, *1:* 74
types of, *1:* 39-40
uses of, *1:* 40
Coal gasification, *1:* 43, **44-46**,
  44 (ill.)
Coal mines
  description of, *1:* 39
  environmental impact of,
    1: 12, 13 (ill.), 40, 42
  fires in, *1:* 41
  restoration of, *1:* 42
Coal stoves, *1:* 4, 40
Coal-burning electric power
  plants
  air pollution from, *1:* 20, 41-42
  uses of, *1:* 8, 11, 40
  water for, *3:* 330
Coal-burning engines, *1:* 6-7
Coast Guard (U.S.), *1:* 19
Cob buildings, *3:* 349-50
Cogeneration, *1:* 35
Coke, petroleum, *1:* 27
Cold fusion, *2:* 189, 191,
  3: 401-5, 403 (ill.), 408
Cold temperature
  biodiesel and, *1:* 78
  vegetable oil fuels and,
    1: 82
Collectors, solar. *See* Solar
  collectors
Colorado
  cost of electricity in, *3:* 333
  hybrid vehicles in, *3:* 369
  wind energy in, *3:* 320, 330

Colorado River dams, *3:* 279

Colter, John, *1:* 98

Columbia River dams, *3:* 279, 282, 283-84

Colville Indians, *3:* 289

Combined cycle generation, *1:* 36

Commercial buildings
climate-responsive, *3:* 346-47
energy consumption by, *3:* 345

Compact fluorescent bulbs, *3:* 357-58, 361 (ill.)

Compass, magnetic, *3:* 387, 389 (ill.)

Compost, *1:* 71-72

Compressed gas tanks, *2:* 161

Compressed natural gas (CNG), *1:* 36

Concentrated solar power systems, *2:* 218

Condensate wells, *1:* 33

Connecticut
cost of electricity in, *3:* 333
hybrid vehicles in, *3:* 369

Conservation of energy. *See* Energy conservation and efficiency

Consumption, reducing, *3:* 374-75

Contests, solar energy, *2:* 232

Controlled nuclear fusion, *3:* 399-401

Convecting solar ponds, *2:* 248-50

Convective loop system, *2:* 225

Conventional nuclear fusion, *3:* 396-97

Cooking
solar, *3:* 362, 363 (ill.)
syngas for, *2:* 139

Cooling
in climate-responsive buildings, *3:* 346
energy conservation and efficiency for, *3:* 359-60
natural gas for, *1:* 35
ocean thermal energy conversion for, *3:* 292
solar collectors for, *2:* 210, 218

solar thermal, *2:* 219

transpired solar collectors for, *2:* 229

Core (Earth), *1:* 105, 105 (ill.)

Coriolis effect, *3:* 270, 316, 317 (ill.)

Coriolis, Gaspard-Gustave de, *3:* 316

Corn
ethanol from, *1:* 91, 92
hydrogen from, *2:* 149

Corporate Average Fuel Economy (CAFE), *1:* 32

Costs
of biodiesel, *1:* 78
of coal gasification, *1:* 46
of electricity, *3:* 333
of hydrogen fuel cells, *2:* 157
of hydropower, *3:* 278
of liquefied petroleum gas, *1:* 50
of nuclear power plants, *2:* 204-6, *3:* 381-82
of nuclear waste, *2:* 206
of petroleum, *1:* 28-29, *3:* 271
of photovoltaic cells, *2:* 240-41
of P-Series fuels, *1:* 93-94
of solar energy, *2:* 219, 221-22, *3:* 383-84
of tidal power, *3:* 271, 298, 299
of uranium, *2:* 194
of wind energy, *3:* 333
of wind turbines, *3:* 333
See also Economic impact

Cree Indians, *3:* 288-89

Crewdson, Eric, *3:* 264

Crops, drying, *1:* 103, 117

Crude oil, refining, *1:* 24

Crust (Earth), *1:* 104, 105 (ill.)

CSR User Facility, *2:* 257

Curie, Marie, *2:* 183

Curie, Pierre, *2:* 183

Curies, *2:* 183

Curtains, insulated, *3:* 360

Cylinders, engine, *1:* 4-5

D

Daimler Chrysler
hybrid vehicles, *3:* 371
hydrogen research by, *2:* 151

Daimler, Gottlieb, *1:* 5 (ill.)

Dairy farms, biogas from, *1:* 62, 85

Dams, hydroelectric. *See* Hydroelectric dams

Darrieus wind turbines, *3:* 321, 322 (ill.)

d'Arsonval, Jacques Arsene, *3:* 267, 290

Day, R. E., *2:* 212

Daylighting, *2:* **217-18, 226-28,** *3:* 345

DC (Direct current), *2:* 243

de Coriolis, Gaspard-Gustave, *3:* 316

de Honnecourt, Villard, *3:* 386

de Saussure, Horace-Benedict, *3:* 343-44

Decomposition
greenhouse gases from, *3:* 287
of manure, *1:* 70

Deforestation, *1:* 74

Dehumidification, *1:* 35

Denmark
biogas in, *1:* 85
wind energy in, *3:* 315, 320

Department of Energy. *See* U.S. Department of Energy

Deposits. *See* Reserves and deposits

Desalination
ocean thermal energy conversion for, *3:* 290, 292
solar ponds for, *2:* 251

Deserts
dish systems in, *2:* 245
trough systems in, *2:* 246, 247-48

Deuterium, *3:* 396-97, 400, 402, 406

Deuteron, *2:* 190-91

Developing nations
coal-burning power plants in, *1:* 42
dish systems for, *2:* 245

fossil fuel use in, *1:* 19, 20
geothermal energy for, *1:* 98, 114
hydrogen energy for, *2:* 164
hydropower in, *3:* 277
nuclear energy for, *2:* 207
solar energy for, *2:* 242-43
Diesel engines
biodiesel for, *1:* 59, 75-80
invention of, *1:* 4
use and workings of, *1:* 6
vegetable oil fuels for, *1:* 80-82
Diesel fuel. *See* Biodiesel; Petrodiesel
Diesel, Rudolf, *1:* 4, 63
Diffuse radiation, *2:* 217
Digestion technology, *1:* 84, 85
Dinosaur fossils, *1:* 3
Direct current (DC), *2:* 243
Direct radiation, *2:* 217
Dirigibles, *2:* 137-39
Dirty bombs, *2:* 200-201
Dish systems, *2:* 213-15, 243-46, 244 (ill.)
Dish-engine system, *2:* 243, 245
Dishwashers, energy efficient, *3:* 362
Distillate fuel oil, *1:* 26
Distributed-point-focus systems. *See* Dish systems
Dogru, Murat, *2:* 149
Domesday Book, 3: 262
Doping agents, *2:* 237
Double-hulled oil tankers, *1:* 19
Double-paned windows, *3:* 360
Drag, blade configuration for, *3:* 324
Drag coefficient, *3:* 369
Drainage, windmills for, *3:* 307-8, 311
Drainback systems, *2:* 232
Draindown systems, *2:* 232
Drake, Edwin, *1:* 10
Dreden-I Nuclear Power Station, *2:* 179
Drilling for geothermal energy, *1:* 107, 111
See also Oil wells
Driving, fuel-efficient, *3:* 372-74

Dry steam geothermal power plants, *1:* 109, 122, 124
Drying
clothes, *3:* 362
crops, *1:* 103, 117
solar energy for, *2:* 210
Dung
air pollution from, *1:* 74
biofuel, *1:* 69-70, 74
Dye factories, *1:* 131

E

E85, *1:* 89-90
Earth (Planet)
Coriolis effect and, *3:* 316, 317 (ill.)
energy footprint on, *3:*374-77
structure of, *1:* 104-5, 105 (ill.)
tides and, *3:* 295-96
wind energy and, *3:* 317-18
Earth bag buildings, *3:* 350
Earthquakes, *1:* 113
Ebb generating systems, *3:* 296-97
Economic impact
of biodiesel, *1:* 67, 78, 79-80
of biogas, *1:* 87
of coal, *1:* 42
of coal gasification, *1:* 46
of daylighting, *2:* 228
of dish systems, *2:* 245
of ethanol, *1:* 67, 91
of fossil fuels, *1:* 19-20, 67
of geothermal energy, *1:* 113, 125, 129
of hybrid vehicles, *3:* 369-70, 371-72
of hydroelectricity, *3:* 274, 287-88
of hydrogen, *2:* 163
of hydrogen fuel cells, *2:* 157
of liquefied petroleum gas, *1:* 49-50
of methanol, *1:* 51
of natural gas, *1:* 37
of nuclear energy, *2:* 204-6
of ocean thermal energy conversion, *3:* 274, 294

of ocean wave power, *3:* 302
of petroleum, *1:* 28-29
of photovoltaic cells, *2:* 241-42
of P-Series fuels, *1:* 93-94
of solar energy, *2:* 219, 221-22, *3:* 383-84
of solar furnaces, *2:* 258
of solar water heating systems, *2:* 235
of tidal power, *3:* 298, 299
of transpired solar collectors, *2:* 230
of vegetable oil fuels, *1:* 83
of water energy, *3:* 271, 274
of wind energy, *3:* 322, 333-34, 335
Ecosystem impact. *See* Environmental impact
Edison, Thomas, *3:* 353
Eel farms, *1:* 119
Efficiency, energy. *See* Energy conservation and efficiency
Egyptians
hydroelectric dams and, *3:* 286
methanol use by, *1:* 50
Sun Gods of, *2:* 209
waterwheels and, *3:* 261
Einstein, Albert, *2:* 173, 177, *3:* 397
Eisenhower, Dwight D., *2:* 178-79
Elastomeric hose, *3:* 300
Electric lights. *See* Lighting
Electric motors, *3:* 365, 366, 370
Electric power plants
air pollution from, *1:* 12
coal-burning, *1:* 8, 11, 20, 40, 41-42, *3:* 330
cogeneration, *1:* 35
combined cycle generation, *1:* 36
for factories, *1:* 11
fossil fuels for, *1:* 7-8
health effects of, *3:* 331
hybrid, *1:* 110, 123
hydrogen-powered, *2:* 165
natural gas, *1:* 8, 37

photovoltaic cell, 2: 240,
3: 383 (ill.)
solar power satellites as,
3: **407-8**
tidal, 3: 264-65
water for, 3: 330
See also Geothermal power
plants; Hydroelectricity;
Nuclear power plants
Electric stoves, 1: 49
Electric vehicles, 3: 341, 365-66
Electricity
for commercial buildings,
3: 345
DC *vs.* AC, 2: 243
hydrogen-powered
generators for, 2: 145 (ill.),
154, 164, 165
magnetism and, 3: **386-90**
per kilowatt-hour cost of,
3: 333
transportation of, 3: 382
Electricity-generating film,
2: 219
Electrolysis
high temperature, 2: 147
for hydrogen production,
2: 133, 147, 148-49, 160
wind-powered, 2: 160
Electromagnetism, 3: 386-90,
387 (ill.)
Electrons, 2: 169, 170, 3: 397
Electrostatic repulsion, 3: 397
Elements
atomic numbers and, 2: 170
periodic table of, 2: 172
Elliot, William Bell, 1: 99
Elugelab Island, 3: 398, 399
(ill.)
Embalming, 1: 50
Embodied energy, 3: 352
E=Mc^2 equation, 2: 173, 3: 397
Emissions. *See* Air pollution;
Exhaust emissions
Encas, Aubrey, 2: 214-15
Endangered species, 2: 203
Energy
alternative, 3: 380-84
definition of, 3: 379
embodied, 3: 352

E=Mc^2 equation for, 2: 173,
3: 397
fake sources of, 3: 384-85
free, 3: 384-85
hydraulic, 3: 268
quantum physics and,
3: 392-93
zero point, 3: 385, **390-95,**
408
See also Future energy
sources; Kinetic energy;
Renewable energy
Energy conservation and
efficiency, 3: 337-77
appliances and, 3: 360-64
definition of, 3: 337-38
energy footprint and,
3: **374-77**
future energy sources and,
3: 408
for homes, 3: **358-60**
for lighting, 3: **352-58**, 360,
361 (ill.), 376
for oil and gas, 3: 338-39
for transportation, 3: **364-65**
See also Climate-responsive
buildings; Green building
materials
Energy footprint, 3: 374-77
Energy Information
Administration, 3: 333
Energy Policy Act (1999), 1: 93
Energy Star Ratings, 3: 364
Enfield-Andreau wind turbine,
3: 315
Engine-block heaters, 3: 373
Engines, 1: 80-82, 2: 150-57
of coal gasification, 1: 45, 46
coal-burning, 1: 6-7
diesel, 1: 4, 6, 59, 75-82
for dish-engine systems,
2: 243
four-stroke, 1: 4-5, 62
idling, 3: 373
jet, 1: 26, 2: 139, 3: 395
knocks, 1: 6, 27, 90
liquefied petroleum gas, 1: 6
natural gas, 1: 6
steam, 1: 6-7, 7 (ill.), 9, 9
(ill.), 10, 71 (ill.)

See also Internal combustion
engines
England
hot springs in, 1: 100
hybrid vehicles in, 3: 370
tidal power in, 3: 297-98,
299
waterwheels in, 3: 262
wind turbines in, 3: 315
windmills in, 3: 307
See also United Kingdom
Eniwetok atoll, 3: 398
Environmental impact
of biodiesel, 1: 65, 78
of bioenergy, 1: 65-67
of biogas, 1: 86-87
of coal, 1: 40-42
of coal mines, 1: 12, 13 (ill.),
30, 42
of daylighting, 2: 228
of dish systems, 2: 245
of ethanol, 1: 90-91
of fossil fuels, 1: 11-19, 20
of gasoline, 1: 28
of geothermal energy,
1: 112-13, 118, 120-21,
124-25, 129
of hydroelectric dams,
3: 273, 286, 289-90
of hydrogen, 2: 162-63
of hydrogen fuel cell
vehicles, 2: 162-63
of liquefied petroleum gas,
1: 49
of methane, 1: 86-87
of methanol, 1: 51-52
of MTBE, 1: 53-54
of natural gas, 1: 37
of nuclear power plants,
2: 196, 202-4
of ocean thermal energy
conversion, 3: 293
of ocean wave power, 3: 301,
302
of petroleum, 1: 28
of photovoltaic cells, 2: 241
of P-Series fuels, 1: 93
of renewable energy, 3: 382
of solar energy, 2: 220-21
of solar furnaces, 2: 258

of solar ponds, *2:* 251-52

of solar water heating systems, *2:* 235

of solid biomass, *1:* 74

of tidal power, *3:* 273, 298

of transpired solar collectors, *2:* 230

of trough systems, *2:* 247-48

of vegetable oil fuels, *1:* 82-83

of water energy, *3:* 271, 273

of wind energy, *3:* 321-22, 323 (ill.), 332-33

See also Pollution

Environmental Protection Agency (EPA)

on acid rain, *1:* 16

on air pollution, *1:* 14

on carbon monoxide levels, *1:* 91

Clean Air Act and, *1:* 15

on hydrogen fuel cells, *2:* 154

on MTBE, *1:* 53

on oil spills, *1:* 19

on sulfur dioxide, *1:* 14

EPA. *See* Environmental Protection Agency

Equality Mills, *3:* 277

Ericsson, John, *2:* 214

Erosion

from ocean waves, *3:* 270

from wind farms, *3:* 332

Erren, Rudolf, *2:* 154

Ethane

from natural gas, *1:* 30, 34, 35

processing, *1:* 47

uses of, *1:* 48

Ethanethiol, *1:* 47

Ethanol, *1:* 60 (ill.), 63, **87-92**

benefits and drawbacks of, *1:* 64, 90

in biodiesel production, *1:* 76

economic impact of, *1:* 67, 91

energy to produce, *1:* 88-89, 91, 92

environmental impact of, *1:* 90-91

in gasohol, *1:* 87-89, 90, 91

history of, *1:* 60

issues and problems with, *1:* 92

vs. methanol, *1:* 51

production of, *1:* 67, 88-89, 91

in P-Series fuels, *1:* 92

sources of, *1:* 72, 90, 92

from sugarcane, *1:* 60, 60 (ill.), 70, 89, 92

uses of, *1:* 60, 89-90

Ethanol poisoning, *1:* 89

Ethyl alcohol. *See* Ethanol

Ethylene, *1:* 48

Europe

biodiesel in, *1:* 76-77, 79

diesel engines in, *1:* 6

syngas in, *2:* 139

windmills in, *3:* 306-9, 309 (ill.)

See also specific countries

Evacuated tube solar collectors, *2:* 234

Exhaust emissions

air pollution from, *1:* 12

Clean Air Act on, *2:* 142-43

from diesel engines, *1:* 6

from P-Series fuels, *1:* 93

Exhaust valves, *1:* 5

Exxon Valdez, *1:* 18, 19

F

Factories

coal power for, *1:* 40

dye, *1:* 131

geothermal energy for, *1:* 110, 130-31

liquefied petroleum gas for, *1:* 48

power generation for, *1:* 11

water energy for, *3:* 262

Faraday, Michael, *3:* 282

Farms

for biofuels, *1:* 66-67

compost for, *1:* 72

dairy, *1:* 62, 85

ethanol production on, *1:* 91, 92

fish, *1:* 110, 118-21

liquefied petroleum gas for, *1:* 48

tree, *1:* 74

wind turbines on, *3:* 330

windmills for, *3:* 312

See also Agriculture; Wind farms

Federal Express fuel cell vehicles, *2:* 152

Federal Wind Energy Program (U.S.), *3:* 316

Feedstock

biofuel, *1:* 66

petroleum, *1:* 27

Fermi, Enrico, *2:* 171-76, 177

Fermi Prize in Physics, *2:* 174

Fertilizer, *1:* 72

FFVs. *See* Flexible fuel vehicles

Fighter jets, *3:* 395

Filament

carbon-based, *3:* 353-54

tungsten, *3:* 354-55

Filling stations, hydrogen, *2:* 146, 159-60

Film, electricity-generating, *2:* 219

Filters, HEPA, *1:* 16

Fires

burning glasses for, *2:* 212

in coal mines, *1:* 41

dung for, *1:* 69-70

Firewood, *3:* 342

Fish

hydroelectric dams and, *3:* 286, 289-90

tidal power and, *3:* 298

Fish farms, *1:* 110, 118-21

Fish ladders, *3:* 286

Fission

nuclear, *2:* 169-84, 190

spontaneous, *2:* 171

Flashed steam geothermal power plants, *1:* 102, 103, 108, 122-23, 124

Flat-plate solar collectors, *2:* 232, 234

Fleet vehicles, *1:* 65, 77

Fleischmann, Martin, *2:* 191, *3:* 402-4, 403 (ill.)

Flexible fuel vehicles (FFVs)

description of, *1:* 65

ethanol for, *1:* 63

light trucks as, *1:* 89

P-Series fuels for, *1:* 92, 93

Flood-generating systems, 3: 296-97
Florida, cost of electricity in, 3: 333
Fluidized bed technology, 1: 43
Fluorescent lightbulbs, 3: 355-58, 360, 361 (ill.)
Food choices, energy efficient, 3: 376-77
Food processing, 1: 103
Ford Focus, 2: 156
Ford, Henry, 1: 10, 60
Ford Motor Company
    fuel cell research by, 2: 156
    hybrid vehicles, 3: 371
Forebay tanks, 3: 277
Formaldehyde, 1: 52, 65
Fossil fuels, 1: 1-55
    air pollution from, 1: 12-15, 20
    benefits and drawbacks of, 1: 11
    bioenergy substitutes for, 1: 57
    biofuels and, 1: 61, 65, 67
    current consumption of, 3: 339
    deposits of, 1: 3-4
    economic impact of, 1: 19-20, 67
    energy conservation and efficiency for, 3: 337
    environmental impact of, 1: 11-19, 20
    for ethanol production, 1: 67, 88-89, 91
    fertilizer from, 1: 72
    history of, 1: 8-10
    how they work, 1: 4-8
    reserves of, 3: 338
    societal impact of, 1: 20
    technology for, 1: 10-11
    types of, 1: 1-4
    water pollution from, 1: 20
    See also Coal; Liquefied petroleum gas; Natural gas; Petroleum
Fossils, dinosaur, 1: 3
Four-stroke engines, 1: 4-5, 62
Fractionation, 1: 34, 47

France
    biodiesel in, 1: 76-77
    biofuels in, 1: 67
    ethanol production in, 1: 60-61
    geothermal heating in, 1: 128
    nuclear fusion research in, 2: 190, 3: 401
    nuclear power plants in, 2: 180, 202
    solar furnaces in, 2: 256-57
    tidal power in, 3: 265, 297
    windmills in, 3: 307
Francis, James B., 3: 281
Francis turbine, 3: 281
Free energy, 3: 384-85
FreedomCAR and Fuel Initiative, 2: 143-44
Frisch, Otto, 2: 174
Fritts, Charles, 2: 213
Fuel additives, 1: 27, 52-54, 90
Fuel cells
    alkaline, 2: 145-46
    for cellular telephones, 1: 52
    ethanol, 1: 90
    methanol, 1: 50, 52, 88 (ill.), 90
    natural gas, 1: 35
    proton exchange membrane, 2: 151
    standards for, 2: 157
    See also Hydrogen fuel cells
Fuel oil, 1: 26
Fuel tanks, 3: 366
Fuel-efficient driving, 3: 372-74
Fuels. See Biofuels; Fossil fuels; Vehicle fuels
Fuller, Calvin, 2: 213
Fulton, Robert, 1: 9
Fumaroles, 1: 108
Furnaces
    natural gas, 1: 4
    propane, 1: 49
    solar, 2: 219, **255-58**, 256 (ill.)
Fusion. See Nuclear fusion
Future energy sources, 3:379-409
    alternative energy as, 3: 380-84

electromagnetism, 3:**386-90**, 387 (ill.)
free energy dreams, 3: 384-85
perpetual motion, 3: 384, **385-86**
zero point energy, 3: 385, **390-95**, 408
See also Nuclear fusion

G

Gamma rays, 2: 186
Garbage
    biogas from, 1: 72, 85
    ethanol from, 1: 91
    P-Series fuels from, 1: 92
Gardening, organic, 1: 72
Gas. See Gasoline; Liquefied petroleum gas; Natural gas
Gas hydrates, 1: 3
Gas mileage, 1: 32, 3: 364-65, 371
Gas tanks, compressed, 2: 161
Gas wells, 1: 33
Gases, greenhouse. See Greenhouse gases
Gasification, coal, 1: 43, 44-46, 44 (ill.)
Gasohol, 1: 81 (ill.), 87-89, 90, 91
Gasoline
    air pollution from, 1: 28
    aviation, 1: 27
    biodiesel with, 1: 75
    butane in, 1: 48
    current consumption of, 3: 364
    environmental impact of, 1: 28
    history of, 1: 10
    leaded, 1: 27
    methanol with, 1: 51
    MTBE in, 1: 27, 50, 52-54
    vs. natural gas, 1: 2
    octane of, 1: 6, 27, 52, 90
    oxygenated, 1: 53
    refining, 1: 24
    reformulated, 1: 53
    uses of, 1: 26
    See also Internal combustion engines

Gearboxes
   self-lubricating, *3:* 311-12
   for waterwheels, *3:* 278
   for windmills, *3:* 308, 311-12
Gedser wind turbine, *3:* 315
Gemini spacecraft, *2:* 139-40, 151
General Electric, fuel cell research by, *2:* 151
General Motors
   hybrid vehicles, *3:* 371
   hydrogen research by, *2:* 151-53, 160
Generators
   in hybrid vehicles, *3:* 367
   for hydroelectricity, *3:* 281-82
   hydrogen fuel cells for, *2:* 145 (ill.), 154, 164, 165
   for ocean wave power, *3:* 300
   for waterwheels, *3:* 278
Genghis Khan, *3:* 306
Geo-Heat Center, *1:* 103
Geologists, *1:* 22
GeoPowering in the West initiative, *1:* 104
Geothermal energy, *1:* 97-132
   for agriculture, *1:* 115-18
   for aquaculture, *1:* 110, 118-21
   barriers to, *1:* 114-15
   benefits and drawbacks of, *1:* 111-12, 117-18, 120
   direct use of, *1:* 110
   economic impact of, *1:* 113, 125, 129
   environmental impact of, *1:* 112-13, 118, 120-21
   history of, *1:* 98-104
   how it works, *1:* 104-6
   industrial applications of, *1:* 130-31
   sites for, *1:* 107-8, 126
   societal impact of, *1:* 113-14
   technology for, *1:* 108-11
   types of, *1:* 97-98
   uses of, *1:* 106-7
Geothermal Energy Association, *1:* 103

Geothermal Energy Research, Development and Demonstration Act, *1:* 103
Geothermal heat pumps, *1:* 126-28, 129
Geothermal heating, *1:*125-30, 127 (ill.)
   benefits and drawbacks of, *1:* 128-29
   direct, *1:* 125-26
   for greenhouses, *1:* 102, 110, 116, 117-18, 117 (ill.)
   history of, *1:* 100, 102
   impact of, *1:* 129
   issues and problems with, *1:* 129-30
Geothermal power plants, *1:* 97-98, 104 (ill.), 109 (ill.), **121-25,** 121 (ill.)
   benefits and drawbacks of, *1:* 111-12, 124
   binary, *1:* 103, 108-9, 122, 123 (ill.)
   dry steam, *1:* 109, 122, 124
   economic impact of, *1:* 125
   environmental impact of, *1:* 112-13, 124-25
   flashed steam, *1:* 102, 103, 108, 122-23, 124
   history of, *1:* 101-4
   hot dry rock, *1:* 103, 111
   hybrid, *1:* 110, 123
   issues and problems with, *1:* 125
   steam, *1:* 101-2
   technology for, *1:* 108-10
Geothermal reservoirs, *1:* 107, 112, 113, 114, 125
Geothermal Resources Council, *1:* 103
Geothermal Steam Act, *1:* 103
Germanium, *2:* 237
Germany
   airships in, *2:* 138-39
   biodiesel in, *1:* 67, 76-77
   hot springs in, *1:* 100
   hydrogen filling stations in, *2:* 159
   hydrogen research in, *2:* 145-46

   nuclear weapons research in, *2:* 176
   solar energy in, *2:* 220
   solar furnaces in, *2:* 256
   wind energy in, *3:* 315, 320
Geysers, *1:* 99-100, 101-2, 108, 122
Glass panels, vision, *3:* 344 (ill.)
Glasses, burning, *2:* 212
Global warming, *1:* 16-18, *3:* 339, 340
Glycerin, *1:* 76
Gods, sun, *2:* 209-10
Government support
   for nuclear power plants, *2:* 179-80
   for solar energy, *2:* 216
Graf Zeppelin (Airship), *2:* 138-39, 140 (ill.)
Grain, ethanol from, *1:* 60, 72
Grain mills
   water-powered, *3:* 262, 277
   wind-powered, *3:* 306, 308
Grand Coulee Dam, *3:* 279, 289
Grand Rapids, Michigan, *3:* 264
Grant, John D., *1:* 101-2
Graphite, *2:* 188
Gravity, *3:* 268-69
Great Depression, *3:* 313
Great Dismal Swamp, *1:* 38
Great Plains, windmills in, *3:* 310-11
Greeks
   climate-responsive buildings by, *3:* 342-43
   fossil fuels for, *1:* 8
   magnetism and, *3:* 387
   solar energy and, *2:* 210
   Sun Gods of, *2:* 209-10
   waterwheels and, *3:* 261
Green building materials, *3:* 341, **347-52**
Greenhouse gases, *1:* 16-18
   from decaying vegetation, *3:* 287
   from methanol, *1:* 51-52
   from natural gas, *1:* 37
   from solid biomass, *1:* 74

Greenhouses
  in climate-responsive
    buildings, 3: 343 (ill.)
  geothermal heating for,
    1: 102, 110, 116, 117-18,
    117 (ill.)
  passive solar design for,
    2: 212, 224
  Roman, 3: 343
Grid intertie, 3: 329
Grills, 1: 62
Grindstones, 3: 308
Ground water pollution,
  1: 53-54
Gulf Stream, 3: 270
Guthrie, Woodie, 3: 282

H

H2R concept car, 2: 155-56,
  158 (ill.)
Habitat degradation. *See*
  Environmental impact
Hahn, Otto, 2: 172-73, 174
Half-life
  of plutonium, 2: 197
  of uranium, 2: 185-86, 197
Halladay, Daniel, 3: 311
Halladay Windmill Company,
  3: 311
Halogen lightbulbs, 3:
  354-55
Handheld windmills, 3: 305
Hawaii
  cost of electricity in, 3: 333
  ocean thermal energy
    conversion in, 3: 267,
    291-92
Hazelnuts, hydrogen from, 2:
  149
Health effects
  of buildings, 3: 347
  of coal mines, 1: 43
  of electric power plants, 3:
    331
  of fossil fuels, 1: 12
  of MTBE, 1: 53
Heat
  from fossil fuels, 1: 4
  in nuclear fusion, 3: 397
  in nuclear reactors, 2: 186

Heat exchangers
  in nuclear power plants, 2: 189
  for solar ponds, 2: 249
  in solar water heating
    systems, 2: 232-33
Heat pumps, geothermal,
  1: 102, 110, 126-28, 129
Heat values, defined, 1: 39
Heating
  in climate-responsive
    buildings, 3: 346
  coal for, 1: 40
  daylighting for, 2: 227
  energy conservation and
    efficiency for, 3: 359-60
  engine-blocks, 3: 373
  fuel oil for, 1: 26
  hypocaust for, 1: 73
  natural gas for, 1: 35
  passive solar design for,
    2: 210-12, 225-26, 3: 342
  propane for, 1: 48, 49
  solar ponds for, 2: 251
  syngas for, 2: 139
  thermal mass and, 3: 351
  transpired solar collectors
    for, 2: 228-30
  See also Geothermal heating;
    Water heaters
Heavy hydrogen, 2: 190-91,
  3: 399-401
Heavy water, 3: 402-3
Heisenberg, Werner, 3: 392-93,
  394
Helios (God), 2: 209-10
Heliostat mirror towers. *See*
  Solar towers
Heliostats, 2: 227, 253, 255
Helium
  nuclear fisson and, 2: 190-91
  nuclear fusion and, 3: 397,
    400, 403, 404
HEPA filters, 1: 16
Hero of Alexandria, 3: 305
High level nuclear waste,
  2: 196-99
High temperature electrolysis
  (HTE), 2: 147
High-flux solar furnaces,
  2: 255, 257

High-head hydroelectric plants,
  3: 282
Hill effect, 3: 328
Hindenburg (Dirigible), 2: 141,
  142 (ill.)
Hiroshima bomb, 3: 398-99
Holland, windmills in, 3: 308
**Home energy conservation and
  efficiency, 3:358-60**
  See also Buildings; Passive
    solar design
Honda
  hybrid vehicles, 3: 371
  hydrogen research by,
    2: 153-54
Honda Civic Hybrid, 3: 371
Honda FCX, 2: 154
Honda Insight, 3: 371
Honnecourt, Villard de,
  3: 386
Hoover Dam, 3: 279, 280 (ill.),
  281 (ill.), 283-84
Horizontal waterwheels,
  3: 276
Horizontal-axis wind turbines,
  3: 325
Hose, elastomeric, 3: 300
Hosepump, 3: 300
Hot air balloons, 2: 136
Hot dry rock electric power
  plants, 1: 103, 111
Hot fusion, 2: 189-91, 3: 408
Hot springs
  cause of, 1: 108
  as geothermal energy, 1: 97
  history of, 1: 98-100, 99
    (ill.), 101 (ill.)
  uses of, 1: 110
  See also Geothermal energy
Hot Springs, Arkansas, 1: 98
Hot tubs, 2: 233
Hot water heating. *See* Water
  heaters
HTE (High temperature
  electrolysis), 2: 147
Hubble Space Telescope, 2: 213
Humidity, 3: 346
Hutters wind turbines, 3: 315
Hybrid electric power plants,
  1: 110, 123

Hybrid vehicles, *1:* 32, *3:* 365-74, 366 (ill.), 368 (ill.)
  benefits of, *3:* 367-70
  components of, *3:* 366-67
  economic impact of, *3:* 369-70, 371-72
  future of, *3:* 371-72
  types of, *3:* 370
Hydraulic energy, *3:* 268
Hydrocarbons
  air pollution from, *1:* 14
  from gasoline burning, *1:* 28
  in natural gas, *1:* 30
  petroleum as, *1:* 21
Hydroelectric dams, *3:* 279-90
  in China, *3:* 285 (ill.), 286, 287 (ill.)
  environmental impact of, *3:* 273, 286, 289-90
  history of, *3:* 263-66, 283-84
  vs. nuclear power plants, *2:* 203
  sites for, *3:* 272, 279
  societal impact of, *3:* 288-89
  turbines for, *3:* 280 (ill.)
  world's largest, *3:* 283
Hydroelectricity, *3:* 261, **279-90**
  benefits and drawbacks of, *3:* 272, 284-87
  economic impact of, *3:* 274, 287-88
  generators for, *3:* 281-82
  history of, *3:* 263-66
  issues and problems with, *3:* 289-90, 380
  reservoirs for, *3:* 273, 286, 288
  technology for, *3:* 280-83
  turbines for, *3:* 280 (ill.), 281
  uses of, *3:* 283-84
  waterwheels for, *3:* 277-78
  world's largest plant for, *3:* 283
  world's smallest plant for, *3:* 284
Hydrogen, *2:* **133-67**
  atmospheric, *2:* 162
  benefits and drawbacks of, *2:* 156-57

from biofuels, *1:* 59
biogas, *1:* 84
distribution of, *2:* 159-60
economic impact of, *2:* 163
environmental impact of, *2:* 162-63
future technology for, *2:* 165
future use of, *3:* 383
gaseous, *2:* 158-59, 160-62
heavy, *2:* 190-91, *3:* 399-401
history of, *2:* 134-41, 140 (ill.), 142 (ill.)
infrastructure for, *2:* 163
for internal combustion engines, *2:* 150-57, 163
leaks, *2:* 162
liquid, *2:* 140, 152 (ill.), 159, 160-62
methane from, *1:* 33
in nuclear fusion, *3:* 397, 399-401
production of, *2:* 133, 147-48, 148 (ill.), 160, 162-63, 165
research in, *2:* 139-47
safety of, *2:* 164-65
societal impact of, *2:* 163-65
storage of, *2:* 160-62
transporting, *2:* 157-59
uses of, *2:* 150-56
See also Hydrogen fuel cells
HydroGen 3, *2:* 152
Hydrogen balloons, *2:* 135 (ill.), 136
Hydrogen Economy Miami Energy Conference, *2:* 141
Hydrogen filling stations, *2:* 146, 159-60
Hydrogen fuel cell vehicles
  buses, *2:* 146, 151, 153 (ill.)
  economic impact of, *2:* 163
  environmental impact of, *2:* 162-63
  filling stations for, *2:* 146, 159-60
  H2R concept car, *2:* 155-56, 158 (ill.)
  history of, *2:* 136-37, 137 (ill.)
  hydrogen storage in, *2:* 161

  research on, *2:* 142-47
  uses of, *2:* 150-57
Hydrogen fuel cells
  vs. batteries, *2:* 138
  benefits and drawbacks of, *2:* 156-57
  economic impact of, *2:* 157, 163
  future use of, *3:* 383
  for generators, *2:* 145 (ill.), 154, 164, 165
  history of, *2:* 136-37, 137 (ill.)
  methanol for, *1:* 51
  for public transportation, *2:* 146
  research in, *2:* 141-47
  societal impact of, *2:* 164
  for spacecraft, *2:* 139-40
  uses of, *2:* 150-57
Hydrogen sulfide, *1:* 84, 113
Hydrolysis, *1:* 94
Hydropower, *3:* **275-79**
  benefits and drawbacks of, *3:* 277-78
  costs of, *3:* 278
  history of, *3:* 261-63, 264 (ill.), 275-76, 276 (ill.)
  small-scale, *3:* 277-78
  uses of, *3:* 277
  See also Hydroelectricity
Hydrosizers, *1:* 43
Hypocaust, *1:* 73

I

Iceland
  geothermal energy in, *1:* 104 (ill.), 107-8
  geothermal heating in, *1:* 110, 127 (ill.), 128
  hydroelectricity in, *3:* 272
  hydrogen in, *2:* 142, 146-47
Icelandic New Energy, *2:* 146
ICS systems, *2:* 233
Idaho, geothermal heating in, *1:* 100, 102
Idling, engine, *3:* 373
IEA (International Energy Agency), *1:* 3

Illinois, cost of electricity in, 3: 333

Incandescent lightbulbs, 3: 353-55

Income tax deductions, for hybrid vehicles, 3: 369

India, wind energy in, 3: 320

Indians, American. *See* Native Americans

Indoor air pollution
HEPA filters for, 1: 16
sick building syndrome from, 3: 347, 348
sources of, 1: 14-15

Industrial applications of geothermal energy, 1: 130-31 *See also* Factories

Industrial revolution, 1: 9, 42, 3: 262

Inertial confinement, 3: 401

Infrared cameras, 2: 211

Injection wells, 1: 108

Insulated curtains, 3: 360

Insulation, 3: 360

Integral collector storage (ICS) systems, 2: 233

Intergovernmental Panel on Climate Change, 3: 338-39

Internal combustion engines
biofuels for, 1: 61 62
gasoline for, 1: 4, 5 (ill.), 10-11
history of, 1: 10
how they work, 2: 155
in hybrid vehicles, 3: 365-66, 367, 370
hydrogen-powered, 2: 150-57, 163
liquefied petroleum gas for, 1: 48-49
methane for, 1: 62
use and workings of, 1: 4-6
vegetable oil fuels for, 1: 63

International Association for Hydrogen Energy, 2: 141

International Energy Agency (IEA), 1: 3

International Partnership for the Hydrogen Economy, 2: 144

International Rivers Network, 3: 289

International Solar Energy Society, 2: 220

International Thermonuclear Experimental Reactor (ITER), 2: 190, 3: 401

Intertie systems, 3: 329

Inverters, photovoltaic cell, 2: 237

Iowa, wind energy in, 3: 319, 320, 329

Iran
nuclear proliferation and, 2: 202
windmills in, 3: 306

Irrigation
geothermal energy and, 1: 115-16
hydropower for, 3: 277

Israel, salt-gradient ponds in, 2: 251

Itaipú Dam, 3: 283

Italy
geothermal energy in, 1: 101, 108, 122
hot springs in, 1: 100
wind energy in, 3: 320

ITER (International Thermonuclear Experimental Reactor), 2: 190, 3: 401

J

James Bay project, 3: 286, 288-89

Japan
atomic bombing of, 2: 178
Hiroshima bomb and, 3: 398-99
hydrogen research in, 2: 144-45
nuclear fusion research in, 3: 405
ocean thermal energy conversion in, 3: 267
solar energy in, 2: 220
solar power satellite research in, 3: 408
wind energy in, 3: 320

Japanese macaques, 1: 102

Jeffries, John, 2: 136

Jet engines
hydrogen for, 2: 139
kerosene for, 1: 26
zero point energy for, 3: 395

Jobs. *See* Labor force

Joules, 3: 268

K

Kansas, wind energy in, 3: 318

Keahole Point project, 3: 291-92

Kentucky, cost of electricity in, 3: 333

Kenya
geothermal energy in, 1: 108, 118
photovoltaic cells in, 2: 242

Kerogen, 1: 4

Kerosene
lamps, 1: 10
uses of, 1: 26
vegetable oil fuels with, 1: 82

Kilowatt-hours, 3: 314

Kinetic energy
hydroelectric dams and, 3: 279
in nuclear reactors, 2: 186
ocean wave power as, 3: 299
water energy as, 3: 268-70, 275
of wind, 3: 318, 324, 326

Klamath Falls, Oregon, 1: 126

Klaproth, Martin, 2: 184

Knocks, engine, 1: 6, 27, 90

Komarechka, Robert, 3: 284

Kyoto protocol, 1: 18

L

La Carrindanga Project, 1: 116

La Rance River, 3: 265, 297

Labor force
for nuclear power plants, 2: 206
for wind energy, 3: 333-34

Lamb effect, 3: 394

Lamb, Willis, 3: 394

Lamps, 1: 9-10

Land
for biofuels, *1:* 66-67
for geothermal energy,
1: 115
for hydroelectricity, *3:* 286
for nuclear power plants,
2: 203
for solar energy projects,
2: 220-21
for solar furnaces, *2:* 258
for solar ponds, *2:* 251
for trough systems,
2: 247-48
for wind energy, *3:* 321-22
for wind farms, *3:* 330,
331-32, 382
Landfills, biogas from, *1:* 85, 91
Land-Installed Marine-Powered
Energy Transformer, *3:* 266,
299-300, 301 (ill.)
Landscape
for energy conservation and
efficiency, *3:* 360
for passive solar design,
2: 223
Landslides, *1:* 113
Langer, Charles, *2:* 137
Larderello, Italy, *1:* 101, 122
Lasers, for nuclear fusion,
3: 401
Lava, *1:* 106
Lavoisier, Antoine-Laurent,
2: 134-35
Laws
for coal mine workers,
1: 42-43
for geothermal energy
development, *1:* 102-3
net metering, *3:* 335-36
of thermodynamics, *3:* 386
See also Clean Air Act
Lead, *1:* 27, 41
Leeghwater, Jan Adriannzoon,
3: 309, 311
Levitation, magnetic, *3:* 389-90,
404 (ill.), 405 (ill.)
Licenses, for nuclear power
plants, *2:* 206
Lieb, Charlie, *1:* 102
Lift, *3:* 324-25

Light shelves, *2:* 226
Light straw buildings,
3: 349-50
Light trucks, *1:* 32, 89
Lightbulbs
fluorescent, *3:* 355-58, 360,
361 (ill.)
halogen, *3:* 354-55
incandescent, *3:* 353-55
Lighting
daylighting for, *2:* 217-18,
226-28, *3:* 345
energy efficient, *3:* **352-58**,
360, 361 (ill.), 376
Lignite, *1:* 39
Lime, *1:* 71, *3:* 354
Limpet 500, *3:* 266, 299-300,
301 (ill.)
Line-focus solar collectors,
2: 246
Liquefaction, coal, *1:* 43
**Liquefied petroleum gas
(LPG),** *1:* 26, **46-50**
benefits and drawbacks of,
1: 48-49
engines, *1:* 6
pipelines, *1:* 47
processing, *1:* 34-35, 47
uses of, *1:* 47-48
Liquid biofuels, *1:* 58
Liquid hydrogen
for spaceflight, *2:* 140, 152
(ill.)
storage of, *2:* 160-62
transporting, *2:* 159
Lithuania, nuclear power
plants in, *2:* 180
Loch Linnhe, *3:* 297
Locomotives
coal-burning, *1:* 9
diesel, *1:* 10
hybrid, *3:* 365
maglev, *3:* 389-90, 390 (ill.)
steam engine, *1:* 6-7, 7 (ill.),
71 (ill.), *3:* 311, 312
Looms, water-powered, *3:*
262-63
Lorenz force, *3:* 388-89
Lorenz, Henrick Antoon, *3:* 388
Los Angeles, California, *1:* 17 (ill.)

Louisiana, geothermal energy
in, *1:* 103
Lowell, Francis Cabot, *3:* 263
Lowell, Massachusetts, *3:* 263,
274
Low-level nuclear waste, *2:* 196
LPG. *See* Liquefied petroleum
gas
Lubrication, *1:* 27
Lumens, *3:* 358
Lunar tides, *3:* 295
Lung diseases, *1:* 43

M

M85, *1:* 51
Macaques, Japanese, *1:* 102
Magdeburg Hemisphere, *3:* 392
(ill.)
Maglev spacecraft, *3:* 391 (ill.)
Maglev trains, *3:* 389-90, 390
(ill.)
Magma, *1:* 106, 108
Magnetic bottle method, *3:* 401
Magnetic compass, *3:* 387, 389
(ill.)
Magnetic fields, *3:* 387-88, 400-
401
Magnetic levitation, *3:* 389-90,
404 (ill.), 405 (ill.)
Magnetic poles, *3:* 388
Magnetism and electricity,
3: **386-90**
Malls, daylighting for, *2:* 228
Manhattan Project, *2:* 176-79
Mantle (Earth), *1:* 104-5, 105
(ill.)
Manual transmission, *3:* 372,
374
Manure
as biofuel, *1:* 69-70
biogas from, *1:* 62, 85
methane from, *1:* 66 (ill.)
Mass media, on nuclear energy,
2: 207
Mass, thermal. *See* Thermal
energy systems
Massachusetts, water energy in,
3: 263, 274
Meckong River, *3:* 273
Medium-Btu gas, *1:* 45

Megawatts, 3: 314

Meitner, Lise, 2: 173, 174

Meltdown, 2: 182

Meltdown at Three Mile Island (Documentary), 2: 207

Membrane ponds, 2: 249-50

Mesopotamians, 1: 8, 2: 210

Methane, 1: 59, 66 (ill.), 84-87
  atmospheric, 1: 17
  biomass sources of, 1: 70
  from coal gasification, 1: 45
  deposits of, 1: 3
  environmental impact of, 1: 86-87
  for internal combustion engines, 1: 62
  making your own, 1: 33
  natural gas and, 1: 30, 33-34, 47
  production of, 1: 33, 47, 86 (ill.)
  sources of, 1: 18

Methanol
  in biodiesel production, 1: 76
  biogas, 1: 87-92
  fossil fuel, 1: 50-52
  fuel cells, 1: 50, 52, 88 (ill.), 90
  in geothermal heat pumps, 1: 129
  from natural gas, 1: 35

Methanol poisoning, 1: 51

MeTHF (Methyltetrahydrofuan), 1: 92, 93, 94

Methyl alcohol. *See* Methanol

Methyl tertiary-butyl ether (MTBE), 1: 27, 50, **52-54**

Methyltetrahydrofuan (MeTHF), 1: 92, 93, 94

Mica windows, 3: 342-43

Michigan, hydroelectricity in, 3: 264

Microbes, for hydrogen production, 2: 149

Microgenerators, 3: 284

Microwaves, 3: 407

Middle East
  oil embargo and, 1: 61, 91, 3: 316

petroleum reserves, 1: 30
  windmills in, 3: 305-6

Mike (Bomb), 3: 398, 399 (ill.)

Mild hybrid vehicles, 3: 370

Military airships, 2: 138

Mill tailings, 2: 198-99

Million Solar Roofs Initiative, 2: 216

Mills
  grain, 3: 262, 277, 306, 308
  post, 3: 308
  saw, 3: 277
  tower, 3: 308

Mines
  strip, 1: 12, 13 (ill.), 39, 42
  subsurface, 1: 12, 39, 41, 42
  uranium, 2: 176, 187, 194, 198-99
  See also Coal mines

Mini-OTEC, 3: 291-92

Minnesota
  ethanol fuels in, 1: 89, 91
  wind energy in, 3: 319

Mirrors
  parabolic, 2: 243, 244 (ill.), 247 (ill.)
  for solar furnaces, 2: 255

MOD-2 windmill, 3: 321

Model T, 1: 10

Model U, 2: 156

Modules, photovoltaic, 2: 236

Mojave Desert, 2: 246

Molten salt, 2: 253, 254

Mond, Ludwig, 2: 137

Monkeys, 1: 102

Mono-alkyl esters, 1: 76

Monopolies, geothermal energy and, 1: 115

Montgolfier, Étienne, 2: 136

Montgolfier, Joseph, 2: 136

Moon, tides and, 3: 294-96

Motors
  electric, 3: 365, 366, 370
  solar-powered, 2: 214

Mouchout, Auguste, 2: 213-14, 215

Mousse, 1: 18

MTBE (Methyl tertiary-butyl ether), 1: 27, 50, **52-54**

Mudpots, 1: 108

Mummies, embalming, 1: 50

Munzo, D. T., 2: 191

Muon-catalyzed fusion, 3: 405

N

Nacelle, 3: 326-27

National Aeronautics and Space Administration (NASA)
  hydrogen fuel cells for, 2: 139-40
  photovoltaic cells and, 2: 213
  wind turbines and, 3: 321

National Highway Traffic Safety Administration, 1: 32

National Renewable Energy Laboratory, 2: 257

National Wind Technology Center, 3: 317

Native Americans, 3: 288-89, 343

Natural Energy Laboratory of Hawaii Authority (NELHA), 3: 267, 291-92

Natural gas, 1: 30-38
  associated, 1: 33
  benefits and drawbacks of, 1: 36
  vs. biogas, 1: 85
  biomass sources of, 1: 70
  compressed, 1: 36
  conservation of, 3: 338-39
  current consumption of, 3: 339
  deposits of, 1: 3, 32
  economic impact of, 1: 37
  electric power plants, 1: 8
  engines, 1: 6
  environmental impact of, 1: 12, 37
  extraction of, 1: 33
  finding, 1: 32-33
  formation of, 1: 2, 21-22, 31-32
  furnaces, 1: 4
  vs. gasoline, 1: 2
  history of, 1: 9-10
  for hydrogen production, 2: 147-48

issues and problems with,
  1: 37-38
lamps, *1:* 9-10
liquefying, *1:* 34-35
processing, *1:* 33-34, 47
in P-Series fuels, *1:* 92
uses of, *1:* 35-36
vehicles, *1:* 35, 36, 37-38
See also Liquefied petroleum
  gas
Natural gas liquids (NGLs),
  1: 34
Navy (U.S.), nuclear-powered
  ships, *2:* 179
Neap tides, *3:* 295
Nebraska, cost of electricity in,
  3: 333
NECAR 1, *2:* 151
NELHA (Natural Energy
  Laboratory of Hawaii
  Authority), *3:* 267, 291-92
Nelson, Willie, *1:* 64 (ill.)
Neon, *3:* 267
Net metering, *3:* 335-36
Netherlands
  biogas in, *1:* 85
  wind energy in, *3:* 307-8,
    307 (ill.), 311, 320
Neutrons
  definition of, *2:* 170, *3:* 396
  in nuclear fisson, *2:* 171
  in nuclear fusion, *3:* 396-97
Nevada
  geothermal energy in, *1:* 98,
    102
  nuclear waste in, *2:* 198, 199
    (ill.), 200-201
New Mexico, wind energy in,
  3: 320
New York, cost of electricity in,
  3: 333
Newton, Isaac, *3:* 294
NGLs (Natural gas liquids),
  1: 34
Nickel catalysts, *1:* 33
Nigeria, solar energy for, *2:* 242
Nile River dams, *3:* 286
Ninevah, *2:* 212
Nitrate, *1:* 71
Nitrogen biogas, *1:* 84

Nitrogen oxides
  in air pollution, *1:* 14
  from biodiesel, *1:* 65
  from coal, *1:* 41
  from commercial buildings,
    3: 345
  from gasoline burning,
    1: 28
  from methane, *1:* 86
  from natural gas, *1:* 37
Nitrous oxide, *1:* 17, 18
Nobel Prize, for Otto Hahn,
  2: 174
Noise pollution
  hydrogen fuel cell vehicles
    and, *2:* 164
  from ocean wave power,
    3: 301
  from wind turbines, *3:* 331
Nonconvecting solar ponds,
  2: 250
Non-renewable resources, *1:* 11
Noria, 3: 261, 263 (ill.), 264
  (ill.)
North America, windmills in,
  3: 309-17
North American Solar
  Challenge, *2:* 232
North Dakota, wind energy for,
  3: 318
North Korea, nuclear
  proliferation and, *2:* 202
North Pole, *3:* 388
Norway, hydroelectricity in,
  3: 272, 288
Nuclear Control Institute,
  2: 199
Nuclear energy, *2:* 169-208
  barriers to, *2:* 207-8
  benefits of, *2:* 192-95
  drawbacks to, *2:* 192,
    195-202
  economic impact of,
    2: 204-6
  environmental impact of,
    2: 196, 202-4
  history of, *2:* 171-84
  how it works, *2:* 184-88
  safety of, *2:* 181-84, 194-95
  societal impact of, *2:* 206-7

Nuclear fission, *2:* 169-74
  for bombs, *3:* 398-99
  vs. fusion, *2:* 190, *3:* 395
  history of, *2:* 171-84
  See also Nuclear power
    plants
Nuclear fuel rods, *2:* 174-76,
  176 (ill.), 188, 189, 197-98
Nuclear fusion, *2:* 189-91,
  3: **395-406,** 396 (ill.)
  for bombs, *3:* 398-99
  cold, *2:* 189-91, *3:* 401-5,
    403 (ill.), 408
  controlled, *3:* 399-401
  conventional, *3:* 396-97
  definition of, *2:* 169
  vs. fission, *2:* 190, *3:* 395
  hot, *2:* 189-91, *3:* 408
  muon-catalyzed, *3:* 405
  sonofusion, *3:* 406, 408
Nuclear power plants
  accidents in, *2:* 181-84, 182
    (ill.), 185 (ill.), 192 (ill.),
    195-96, 207, *3:* 381
  construction of, *2:* 187-88
  costs of, *2:* 204-6, *3:* 381-82
  decommissioning, *2:* 184, 198
  development of, *2:* 179-84,
    181 (ill.)
  environmental impact of,
    2: 196, 202-4
  history of, *3:* 384
  how they work, *2:* 184-88
  issues and problems with,
    3: 381-82
  labor force for, *2:* 206
  occupational injuries in,
    2: 196
  output of, *2:* 193
  radiation exposure from,
    2: 195
  safety of, *2:* 194-95
  security at, *2:* 201
  technology for, *2:* 188-91
  terrorist attacks and, *2:*
    199-203
  waste from, *2:* 184, 196-99
  water for, *3:* 330
Nuclear reactors, *2:* 176, 181
  (ill.), 184-89, 185 (ill.)

Nuclear Regulatory
  Commission, 2: 206
Nuclear waste
  costs of, 2: 206
  high level, 2: 196-99
  low-level, 2: 196
  storage and disposal of,
    2: 184, 196-99, 199 (ill.),
    200-201
Nuclear Waste Disposal Act,
  2: 198
Nuclear weapons
  development of, 2: 176-79
  nuclear fusion for, 3:
    398-99, 399 (ill.)
  proliferation of, 2: 202
  terrorists and, 2: 200-201
Nuclear-powered submarines,
  2: 179, 180 (ill.)

O

Occupational health
  coal mines and, 1: 43
  nuclear power plants and,
    2: 196
Ocean currents, 3: 264, 270
Ocean thermal energy
  conversion, 2: 250,
  3: 290-94, 291 (ill.)
  benefits and drawbacks of,
    3: 292-93
  economic impact of, 3: 274,
    294
  environmental impact of,
    3: 293
  history of, 3: 266-67
  issues and problems with,
    3: 294
  technology for, 3: 270,
    290-91
Ocean wave power, 3: 299-302
  benefits and drawbacks of,
    3: 272, 300-301
  environmental impact of,
    3: 301, 302
  history of, 3: 264-66
  issues and problems with,
    3: 381
  technology for, 3: 269-70,
    299-300

Oceans
  level of, 1: 18
  size of, 3: 274
  solar energy absorbed by,
    2: 217
  temperature of, 3: 291 (ill.)
Octane, 1: 6
  ethanol and, 1: 90
  MTBE and, 1: 27, 52
  P-Series fuels and, 1: 92
Oersted, Hans Christian, 3: 387
  (ill.)
Oil. See Petroleum; Vegetable
  oil fuels
Oil embargo (1973), 1: 61, 91,
  3: 316
Oil refineries, 1: 24, 26 (ill.)
Oil sands, 1: 25
Oil shale, 1: 4
Oil spills, 1: 18-19, 28
Oil tankers, double hulled, 1: 19
Oil wells, 1: 22-24, 23 (ill.), 29,
  31 (ill.)
  environmental impact of,
    1: 12, 28
  history of, 1: 10
  natural gas from, 1: 33
Oldsmobile, 1: 10
OPEC (Organization of
  Petroleum Exporting
  Countries), 1: 30, 61
Open field agriculture,
  1: 115-16
Open-cycle systems, 3: 290-91
Oppenheimer, J. Robert, 2: 178
Oregon
  geothermal energy in,
    1: 102, 126
  hybrid vehicles in, 3: 369
Organic gardening, 1: 72
Organization of Petroleum
  Exporting Countries
  (OPEC), 1: 30, 61
Organs, wind-powered, 3: 305
Oscillating water column
  (OWC), 3: 300
Oserian Development
  Company, 1: 118
OTEC. See Ocean thermal
  energy conversion

Ott, Nikolaus August, 1: 4
Otto engines, 1: 4-5
Overburden, 1: 39
Overshot waterwheels,
  3: 276-77
OWC (Oscillating water
  column), 3: 300
Ownership rights, water energy
  and, 3: 273, 278-79
Oxygen, 1: 34
Oxygenated gasoline, 1: 53
Oyster Creek Nuclear Power
  Plant, 2: 179
Ozone, 1: 13-14

P

Pacific Ocean Ring of Fire,
  1: 107
Paints, solar, 2: 218-19
Palladium, 3: 402, 403 (ill.)
Palm oil, 1: 75
Palmer, Paul, 2: 191, 3: 402
Parabolic concentrators, 2: 256
Parabolic mirrors, 2: 243, 244
  (ill.), 247 (ill.)
Parabolic solar collectors,
  2: 219
Paraffin wax, 1: 27
Parallel hybrid vehicles, 3: 370
Parana River, 3: 283
Paris, France, 1: 128
Particles (Physics), 3: 392-93
Particulate matter air pollution,
  1: 12-13, 14, 16
Passive lighting. See
  Daylighting
Passive solar design, 2: 222-26,
  223 (ill.)
  convective loop system,
    2: 225
  direct gain, 2: 223
  for greenhouses, 2: 212, 224
  for heating, 2: 210-12,
    225-26, 3: 342
  history of, 2: 210-12
  impact of, 2: 225-26
  roof ponds, 2: 224-25
  technology for, 2: 217-18
  thermal storage, 2: 224
  Trombé walls for, 3: 346

types of, *2:* 223-25
for water heating, *2:* 210-12, 234-35
Paul, Stephen, *1:* 93
PBS (Public Broadcasting Service), *2:* 207
Peat, *1:* 38
Pelton, Lester Allan, *3:* 264, 266
Pelton wheel, *3:* 264, 266
PEM fuel cells, *2:* 151
Penstock, *3:* 277
Pentanes-plus, *1:* 92
Periodic table of the elements, *2:* 172
Permits, for pollutants, *1:* 15
Perpetual motion, *3:* 384, 385-86
Perry, Thomas, *3:* 311
Persia, windmills in, *3:* 305-6
Person, Gerald, *2:* 213
Petrodiesel
biodiesel and, *1:* 63, 77, 78
history of, *1:* 10
ultra-low-sulfur, *1:* 77
vegetable oil fuels with, *1:* 82
Petroleum, *1:* 2, 20-30
benefits and drawbacks of, *1:* 27-29
conservation of, *3:* 338-39
costs of, *1:* 28-29, *3:* 271
current consumption of, *1:* 19, *3:* 339
deposits of, *1:* 3
economic impact of, *1:* 19, 28-29
environmental impact of, *1:* 27
extraction of, *1:* 22-24, 23 (ill.)
feedstock, *1:* 27
finding, *1:* 22
formation of, *1:* 2, 21-22, 24
history of, *1:* 8, 10
issues and problems with, *1:* 29-30
processing, *1:* 24, 26 (ill.)
reserves of, *1:* 28, 29-30, *2:* 194, *3:* 339
uses of, *1:* 25-27

Petroleum coke, *1:* 27
Petroleum industry, *2:* 163
Phantom loads, *3:* 359
Phosphoric acid, *2:* 145
Photoelectric cells. *See* Photovoltaic cells
Photoelectric effect, *2:* 212
Photosphere, *2:* 216-17
Photosynthesis, *2:* 209
Photovoltaic cells, *2:* 236-43, 240 (ill.)
batteries for, *2:* 237-38
benefits and drawbacks of, *2:* 241
building integrated, *2:* 240
in climate-responsive buildings, *3:* 345
economic impact of, *2:* 241-42
efficiency of, *2:* 236, 243
for electric power plants, *2:* 240, *3:* 383 (ill.)
environmental impact of, *2:* 241
future use of, *3:* 382-83
history of, *2:* 212-13
issues and problems with, *2:* 243
polycrystalline, *2:* 237
satellites and, *2:* 213, 238, *3:* 407-8
societal impact of, *2:* 242-43
technology for, *2:* 218
thin-film, *2:* 237-38
types of, *2:* 237-38
uses of, *2:* 238-41
Photovoltaic effect, *2:* 212
Physics, quantum, *3:* 391-93
Pioneer Development Company, *1:* 102
Pipelines
biogas, *1:* 62
hydrogen, *2:* 158
liquefied petroleum gas, *1:* 47
Pistons, *1:* 5, *2:* 155
Pitchblende, *2:* 184
Planck, Max, *2:* 174
Plant-based food products, *3:* 376-77

Plants
aquatic, *1:* 120
ethanol from, *1:* 72
Plasma, in nuclear fusion, *3:* 400-401
Plastics, electricity-generating film on, *2:* 219
Plate tectonics, *1:* 106
Platinum, *2:* 165
Pliny the Younger, *2:* 210
Plug-in hybrid vehicles, *3:* 370
Plumbers, *3:* 361
Plutonium
half-life of, *2:* 197
in nuclear reactors, *2:* 186-87
processing, *2:* 193
terrorists and, *2:* 200-201
weapons-grade, *2:* 205 (ill.)
Point-of-use water heaters, *3:* 362
Poisoning
carbon monoxide, *1:* 15
ethanol, *1:* 89
methanol, *1:* 51
Polar bears, *2:* 211, 211 (ill.)
Pollutants, permits for, *1:* 15
Pollution
noise, *2:* 164, *3:* 301, 331
visual, *3:* 322, 332
See also Air pollution; Water pollution
Polycrystalline photovoltaic cells, *2:* 237
Polymers, *2:* 219
Ponds
membrane, *2:* 249-50
roof, *2:* 224-25
salt-gradient, *2:* 248-49, 249 (ill.), 251, 252
solar, *2:* **248-53**, 249 (ill.)
Pons, Stanley, *2:* 191, *3:* 402-4, 403 (ill.)
Pool systems, swimming, *2:* 233
Porsche, Ferdinand, *3:* 365
Post mills, *3:* 308
Power lines, *3:* 407
Power plants. *See* Electric power plants
Power towers. *See* Solar towers

Undershot waterwheels, 3: 276-77
United Kingdom
  biodiesel in, 1: 79
  ocean wave power in, 3: 266
  wind energy in, 3: 320
  See also England; Scotland
United States
  biodiesel in, 1: 77
  biogas in, 1: 85
  coal in, 1: 41-42
  cost of electricity in, 3: 333
  diesel fuel in, 1: 6
  ethanol in, 1: 67, 89, 91
  gasoline use by, 3: 364
  geothermal energy in, 1: 101-4
  hybrid vehicles in, 3: 369
  hydroelectric dams in, 3: 286, 288
  hydroelectricity in, 3: 264-65, 272, 283 84
  hydrogen research in, 2: 142-44
  nuclear power in, 2: 180, 181-82
  petroleum use in, 1: 28-29
  solar energy in, 2: 242-43
  wind energy in, 3: 318-20, 329
  See also specific states
Uranium
  costs of, 2: 194
  half-life of, 2: 185-86, 197
  isotopes, 2: 172, 186
  mill tailings, 2: 198-99
  mines, 2: 176, 187, 194, 198-99
  in nuclear fuel rods, 2: 174-76, 176 (ill.)
  for nuclear reactors, 2: 184-86, 204
  properties of, 2: 170, 171, 172, 185-86
  sources of, 2: 193
  terrorists and, 2: 200-201
Uranium dioxide, 2: 187
Uranium oxide, 2: 187
U.S. Bureau of Labor Statistics, 2: 196

U.S. Coast Guard, 1: 19
U.S. Department of Energy
  on coal, 3: 340
  GeoPowering in the West Initiative, 1: 104
  on geothermal energy, 1: 103
  on hydrogen filling stations, 2: 159
  Million Solar Roofs Initiative, 2: 216
  on natural gas power plants, 1: 8
  on nuclear fusion, 3: 405
  on nuclear waste, 2: 197, 198
  on propane vehicles, 1: 49
  on transpired solar collectors, 2: 228
  on wind energy, 3: 318, 329
U.S. Federal Wind Energy Program, 3: 316
U.S. Navy, nuclear-powered ships, 2: 179
U.S. Wind Engine and Pump Company, 3: 311
U.S.S. Enterprise (Ship), 2: 179
Utah, geothermal energy in, 1: 98

V

Vacuum, zero point energy and, 3: 390-95
Valves, exhaust, 1: 5
Vanguard 1, 2: 213
Vegetable oil fuels, 1: 80-84
  benefits and drawbacks of, 1: 82
  biodiesel as, 1: 63, 75-76
  economic impact of, 1: 83
  environmental impact of, 1: 82-83
  history of, 1: 59
  issues and problems with, 1: 84
  straight, 1: 78, 80-84
  uses of, 1: 80-82
  from waste oil, 1: 67, 78, 80-84, 83 (ill.)
  See also Biodiesel

Vegetation, decaying, 3: 287
Vehicle fuels
  biofuel, 1: 60-61, 63
  ethanol, 1: 87-88, 89-90
  gasohol, 1: 81 (ill.), 87-89, 90, 91
  liquefied petroleum gas, 1: 50
  methanol, 1: 51
  natural gas, 1: 35, 36, 37-38
  propane, 1: 48-49
  P-Series fuels, 1: 63, 72, 92-94
  types of, 1: 26-27
  See also Biodiesel; Gasoline; Vegetable oil fuels
Vehicles
  electric, 3: 341, 365-66
  energy conservation and efficiency for, 3: 364-65
  exhaust emissions from, 1: 6, 12, 93, 2: 142-43
  fleet, 1: 65, 77
  flexible fuel, 1: 63, 65, 89, 92, 93
  gas mileage of, 1: 32, 3: 364-65, 371
  tips for fuel-efficient driving, 3: 372-74
  See also Automobiles; Hybrid vehicles; Hydrogen fuel cell vehicles
Ventilation systems, 3: 346, 347
Vermont, wind turbines in, 3: 314-15
Verne, Jules, 3: 267
Vertical waterwheels, 3: 276-77
Vertical-axis wind turbines, 3: 321, 325, 384
Virtual particles, 3: 393
Vision glass panels, 3: 344 (ill.)
Visual pollution, 3: 322, 332
Vitrification, 2: 198
Vitruvius, 2: 211-12
Volcanos
  geothermal energy and, 1: 106, 107-8
  hydrogen from, 2: 150
Von Guericke, Otto, 3: 392 (ill.)

W

Wake (Wind energy), 3: 328
Wall transformers, 3: 359
Waltham, Massachusetts, 3: 263
Washing machines, energy efficient, 3: 362
Washington (State), hydroelectricity in, 3: 279, 289
Washington, D.C., hydrogen filling stations, 2: 160
Waste vegetable oil (WVO) fuels, 1: 67, 78, 80-84, 83 (ill.)
Water
    for coal gasification, 1: 46
    for coal-burning electric power plants, 3: 330
    desalinized, 2: 250, 3: 290, 292
    for electric power plants, 3: 330
    from ethanol, 1: 91
    heavy, 3: 402-3
    hydrogen production and, 2: 133, 147, 162-63
    for nuclear power plants, 3: 330
    for steam engines, 3: 311, 312
Water consumption, reducing, 3: 348
Water energy, 3: 261-303
    benefits and drawbacks of, 3: 271-73
    economic impact of, 3: 271, 274
    environmental impact of, 3: 271, 273
    history of, 3: 261-67, 263 (ill.), 264 (ill.)
    how it works, 3: 267-69
    issues and problems with, 3: 382
    ownership rights and, 3: 273, 278-79
    societal impact of, 3: 275
    technology for, 3: 269-70
    See also Hydroelectricity; Hydropower; Ocean

thermal energy conversion; Ocean wave power; Tidal power
Water heaters
    energy conservation and efficiency for, 3: 359
    point-of-use, 3: 362
    See also Solar water heating systems
Water pollution
    from aquaculture, 1: 121
    from fossil fuels, 1: 20
    from MTBE, 1: 53-54
    from nuclear power plants, 2: 202-3
    from tidal power, 3: 298
Water pumps
    hydropower for, 3: 277
    solar-powered, 2: 215
    windmills for, 3: 307-8, 311, 312
Water quality, dams and, 3: 286
Waterwheels, 3: 263 (ill.), 275-79
    history of, 3: 261-63, 264 (ill.), 275-76, 276 (ill.)
    horizontal, 3: 276
    vertical, 3: 276-77
Watt, James, 1: 9, 3: 314
Watts, 3: 314, 355-56
Wave power. See Ocean wave power
WaveEnergy (Company), 3: 300
Wavegen, 3: 299-300
Waves (Physics), 3: 392-93
Wave-surge focusing devices, 3: 300
Wax, paraffin, 1: 27
Weapons-grade plutonium, 2: 205 (ill.)
Wells
    condensate, 1: 33
    gas, 1: 33
    for geothermal energy, 1: 107, 111
    injection, 1: 108
    See also Oil wells
West Virginia, waterwheels in, 3: 277

Wetlands, 2: 203
Whale oil, 1: 59
Wildlife
    nuclear power plants and, 2: 203
    solar energy and, 2: 220-21
    See also Environmental impact
Willsie, Henry, 2: 215
Wind
    availability of, 3: 317-18, 319 (ill.)
    cause of, 3: 317-18
    factors and characteristics, 3: 327-29
    kinetic energy of, 3: 318, 324, 326
    obstacles to, 3: 328-29
Wind energy, 3: 305-35
    benefits and drawbacks of, 3: 321
    currently produced, 3: 319-20
    economic impact of, 3: 322, 333-34, 335
    environmental impact of, 3: 321-22, 323 (ill.)
    history of, 3: 305-17
    how it works, 3: 317-21
    issues and problems with, 3: 382
    mathematics of, 3: 326
    potential for, 3: 318-19, 329
    sites for, 3: 318-21, 319 (ill.)
    societal impact of, 3: 323
    tax credits for, 3: 335
    technology for, 3: 321
Wind Energy Potential, 3: 319 (ill.)
Wind farms, 3: 315 (ill.), 320 (ill.)
    birds and, 3: 332-33
    currently operating, 3: 319-20
    drawbacks to, 3: 331-32
    issues and problems with, 3: 381, 384
    land for, 3: 330, 331-32, 382
    vs. nuclear power plants, 2: 193

uses of, 3: 329-30
visual pollution from, 3: 322
Wind generators. See Wind turbines
Wind patterns, 3: 316
Wind shear, 3: 329
Wind speed, 3: 325, 326, 327, 328
Wind turbines, 3: 315 (ill.), 324-29
benefits of, 3: 330-31
blade configuration for, 3: 321, 324, 326, 327
Darrieus, 3: 321, 322 (ill.)
drawbacks to, 3: 331-32
economic impact of, 3: 333-34
environmental impact of, 3: 332-33
history of, 3: 313-15
horizontal-axis, 3: 325
how they work, 3: 317-21
issues and problems with, 3: 334-35
mathematics of, 3: 326
selecting a location for, 3: 327-29, 382
societal impact of, 3: 334
towers for, 3: 327
uses of, 3: 329-30
vertical-axis, 3: 321, 325, 384
Windbreaks, 3: 360
Windmills

for drainage, 3: 307-8, 311
electricity-generating, 3: 313
in Europe, 3: 306-9, 309 (ill.)
gearboxes for, 3: 308
handheld, 3: 305
history of, 3: 305-13, 307 (ill.), 310 (ill.)
MOD-2, 3: 321
for water pumps, 3: 307-8, 311, 312
Windows
clerestory, 2: 226
coatings and glazings for, 2: 226-27
double-paned, 3: 360
mica, 3: 342-43
Wind-powered electrolysis, 2: 160
Wings, airplane, 3: 324-25
Wisconsin, hydroelectricity in, 3: 264
Wolverine Chair Factory, 3: 264
Wood
biofuel, 1: 69, 73, 74-75
charcoal from, 1: 70-71
shortage of, 3: 342
Wood stoves, 1: 62, 73, 74-75
World Bank
on hydroelectric dams, 3: 289-90
on photovoltaic cells, 2: 242
World Energy Council, 3: 339

World Health Organization
on burning dung, 1: 74
on sick buildings, 3: 347
World Solar Challenge, 2: 232
World War II, 2: 176-79
Wright, Orville, 1: 10
Wright, Wilbur, 1: 10
WVO (Waste vegetable oil) fuels, 1: 67, 78, 80-84, 83 (ill.)
Wyoming, wind energy in, 3: 319

Y

Yangtze River dam, 3: 285 (ill.), 286, 287 (ill.)
Yankee Rowe Nuclear Power Station, 2: 179
Yellowcake, 2: 187
Yellowstone National Park hot springs, 1: 98
Yucca Mountain, 2: 198, 199 (ill.), 200-201

Z

Zambia, geothermal energy in, 1: 108
Zeppelin, Ferdinand Adolf August Heinrich, 2: 138
Zeppelins, 2: 138
Zero point energy, 3: 385, 390-95, 408